Timber
Its nature and behaviour

Timber

Its nature and behaviour

Second edition

J. M. Dinwoodie, *OBE*

Building Research Establishment, and
Honorary Professor, University of Wales

London and New York

with the support of the Centre for Timber Technology and Construction at BRE

First published 1981 by Van Nostrand Reinhold Co. Ltd
Second edition published 2000 by E & FN Spon
11 New Fetter Lane, London EC4P 4EE

Simultaneously published in the USA and Canada
by E & FN Spon
29 West 35th Street, New York, NY 10001

E & FN Spon is an imprint of the Taylor & Francis Group

The first edition of this book comprised the timber sections of *Concrete, Timber and Metals: the nature and behaviour of structural materials* by Illston, Dinwoodie and Smith, © 1979 Van Nostrand Reinhold Co. Ltd, with minor revisions. This second edition is the result of major revision of the previous edition.

Typeset in Times by Florence Production Limited, Stoodleigh, Devon
Printed and bound in Great Britain by St Edmundsbury Press, Bury St Edmunds, Suffolk

British Library Cataloguing in Publication Data
A catalogue record for this book is available from the British Library

Library of Congress Cataloging in Publication Data
A catalogue record for this book is available from the Library of Congress

ISBN 0–419–25550–8 (hbk)
ISBN 0–419–23580–9 (pbk)

Contents

Preface

The initial contents of this text started life in 1979 as a series of chapters in *Concrete, Timber and Metals: the nature and behaviour of structural materials* by J. M. Illston, J. M. Dinwoodie and A. A. Smith. The text was aimed primarily at undergraduate students in structural engineering and material science, who of necessity must study a wide range of materials.

In 1981, the chapters on timber with a few additions were published as a separate text with the title *Timber – its Nature and Behaviour* by J. M. Dinwoodie; this text was aimed primarily at the undergraduate and new post-graduate student in wood science and technology, as well as forestry students with a timber interest, and material science students with a biological interest.

This second edition represents a major revision of the initial text, necessitating in many places the considerable updating of the original text together with the inclusion of much new text embracing new concepts, new explanations, new developments, new techniques and new references.

Students studying timber and coming to the text for the first time will immediately appreciate the marked difference in approach to timber compared with the conventional treatment of the subject in other texts. Although many text-books are available describing, on the one hand, the anatomy and biodeterioration of wood and, on the other hand, the engineering properties and processing of wood, no single text has been previously available to examine comprehensively the relationship between the performance of timber and its basic anatomical and chemical characteristics. Thus, for example, the flow of vapour or liquid water through wood, the movement of wood due both to temperature and moisture change, the stiffness, toughness and strength of wood are explained fully in terms of the basic structure of wood defined at four levels of organisation.

A text of this length cannot claim to be totally comprehensive and certain topics may have received less coverage than others; nevertheless, it is hoped that this text with its limitations will have the interest and content to lead students to a fuller understanding of the nature and behaviour of timber.

J. M. DINWOODIE
Princes Risborough, 1999

Acknowledgements

I wish to express my appreciation to the Building Research Establishment (BRE) and in particular to Dr A. F. Bravery, Director of the Centre for Timber Technology and Construction, not only for permission to use many plates and figures from the BRE collection, but also for providing laboratory support for the production of the figures from both existing negatives and from new material.

Thanks are also due to several publishers for permission to reproduce figures here.

To the many colleagues who have so willingly helped me in some form or other in the production of this revised text I would like to record my very grateful thanks. In particular, I would like to record my appreciation to the following: Dr P. W. Bonfield, BRE, for reviewing the entire text in his own time and in providing the most valuable advice; Dr J. A. Petty, University of Aberdeen, and Dr Hilary Derbyshire, BRE, for their detailed appraisal of Chapter 5 and their most relevant and constructive comments on it; Dr D. G. Hunt, University of the South Bank, for reading Sections 6.3.1.7 and 6.3.1.8 and making valuable detailed comments and proposals; Dr C. Hill, University of Wales, Bangor, for his advice on Section 9.3.1.3; and to the following colleagues at BRE for assistance on specific topics – C. A. Benham, J. Boxall, Dr J. K. Carey, Dr V. Enjily, C. Holland, J. S. Mundy, Dr R. J. Orsler, J. F. Russell, E. D. Suttie and P. P. White. Lastly, my deep appreciation to both my daughter, who did much of the word processing as well as acting as my PC tutor, and my wife, not only for her willing assistance in editing my text and in proofreading, but also for her support, tolerance and patience over many long months.

J. M. D.

Chapter 1

Structure of timber

1.1 Introduction: timber as a material

From earliest recorded times timber has been an ubiquitous material; the ancient Egyptians produced furniture, sculptures, coffins and death masks from it as early as 2500 BC; elaborate wooden couches and beds were produced in the days of the Greek empire (700 BC). The Ancient Briton, somewhat less sophisticated in his requirements, used wood for the handles of his weapons and tools and for the construction of his huts and rough canoes. Considerably more diversity in utilisation appeared in Medieval times when, in addition to the use of timber for the longbow, and later the butt of the crossbow and the chassis of the cannon, timber found widespread use in timber-frame housing and boats; musical instrument manufacture based on wood advanced significantly during this period.

In the industrial era of the nineteenth century timber was used widely for the construction not only of roofs but also of furniture, waterwheels, gearwheels, rails of early pit railways, sleepers, signal poles, bobbins and boats. The twentieth century has seen an extension of its use in certain areas and a decline in others, due to its replacement by newer materials. Despite competition from the lightweight metals and plastics, whether foamed or reinforced, timber continues to be used on a massive scale.

World production of timber in 1993 (the last year for which complete data is available) was 3400×10^6 m^3. As much as 55% of this volume (1880×10^6 m^3) is used as fuelwood, and only 1520×10^6 m^3 is used for industrial and constructional purposes.

In 1997, the UK consumed 48.3×10^6 m^3 of timber, panels, paper and pulp based on an underbark wood raw material equivalent basis (Anon, 1998); this was equivalent to a per capita annual consumption of 0.82 m^3. Consumption of timber and wood-based panels on a wood raw material basis was 27.07×10^6 m^3 comprising 17.48×10^6 m^3 of softwood, 1.86×10^6 m^3 of hardwood, and 7.73×10^6 m^3 of wood-based panels.

About 80% of consumption is met by imports; the cost of these for timber and wood-based panels (after deduction of a small volume of re-exports) was

£4843 million, a far from insignificant import bill and one that is more than 15% of the total UK annual trade deficit.

The remaining 20% of consumption is met from home production, equivalent to 5.51×10^6 m^3 with a value of about £900 million. Although this contribution will increase steadily until about 2020, it is then expected to peak at a value corresponding to only 25% of expected consumption.

In the UK, timber and timber products are consumed by a large range of industries, but the bulk of the material continues to be used in construction, either structurally, such as roof trusses or floor joists (about 43% of total consumption), or non-structurally, such as doors, window frames, skirting boards and external cladding (about 9% of total consumption). The construction industry, therefore, consumed in 1997 timber and wood-based panels to a value of about £3000 million. On a volume basis, annual consumption continues to increase slightly and there is no reason to doubt that this trend will be maintained in the future, especially with the demand for more houses, the increasing price of plastics, the favourable strength–weight and strength–cost factors of timber and panel products, and the increased emphasis on environmental performance and sustainability in which timber is the only renewable construction material.

Timber is cut and machined from trees, themselves the product of nature and time. The structure of the timber of trees has evolved through millions of years to provide a most efficient system which will support the crown, conduct mineral solutions and store food material. As there are approximately 30 000 different species of tree, it is not surprising to find that timber is an extremely variable material. A quick mental comparison of the colour, texture and density of a piece of balsa and a piece of lignum vitae, used to make playing bowls, will illustrate the wide range that occurs. Nevertheless, man has found timber to be a cheap and effective material and, as we have seen, continues to use it in vast quantities. However, he must never forget that the methods by which he utilises this product are quite different from the purpose that nature intended and many of the criticisms levelled at timber as a material are a consequence of man's use or misuse of nature's product. Unlike so many other materials, especially those used in the construction industry, timber cannot be manufactured to a particular specification. Instead the best use has to be made of the material already produced, though it is possible from the wide range available to select timbers with the most desirable range of properties. Timber as a material can be defined as a low-density, cellular, polymeric composite, and as such does not conveniently fall into any one class of material, rather tending to overlap a number of classes. In terms of its high strength performance and low cost, timber remains the world's most successful fibre composite.

Four orders of structural variation can be recognised – macroscopic, microscopic, ultrastructural and molecular – and in subsequent chapters the various physical and mechanical properties of timber will be related to these four levels of structure. In seeking correlations between performance and structure it is tempting to describe the latter in terms of smaller and smaller structural units. Although this desire for refinement is to be encouraged, a cautionary note must be recorded, for it is all too easy to overlook the significance of the gross features. This is particularly so where large sections of timber are being used under practical conditions; in these situations gross features such as knots and grain angle are highly significant factors in reducing performance.

1.2 Structure of timber

1.2.1 Structure at the macroscopic level

The trunk of a tree has three physical functions to perform: first, it must support the crown, the region responsible for the production not only of food but also of seed; second, it must conduct the mineral solutions absorbed by the roots upwards to the crown; lastly, it must store manufactured food (carbohydrates) until required. As described in detail later, these tasks are performed by different types of cell.

Whereas the entire cross-section of the trunk fulfils the function of support, and increasing crown diameter is matched with increasing diameter of the trunk, conduction and storage are restricted to the outer region of the trunk. This zone is known as *sapwood*, whereas the region in which the cells no longer fulfil these tasks is termed the *heartwood*. The width of sapwood varies widely with species, rate of growth and age of the tree. Thus, with the exception of very young trees, the sapwood can represent from 10% to 60% of the total radius, though values from 20% to 50% are more common (Figures. 1.1 and 1.2); in very young trees, the sapwood will extend across the whole radius.

The advancement of the heartwood to include former sapwood cells results in a number of changes, primarily chemical in nature. The acidity of the wood increases slightly, though certain timbers have heartwood of very high acidity. Substances, collectively called *extractives*, are formed in small quantities and these impart not only colouration to the heartwood, but also resistance to both fungal and insect attack. Different substances are found in different species of wood and some timbers are devoid of them altogether. This explains the very wide range in the natural durability of wood, as discussed further in Section 8.3.1. Many timbers develop gums and resins in the heartwood while the moisture content of the heartwood of most timbers is appreciably lower than that of the sapwood in the freshly felled state. However, in exceptional cases high moisture contents can occur in certain parts of the heartwood. Known as *wetwood*, these zones are frequently of a darker colour than the remainder of the heartwood and are thought to be due to the presence of micro-organisms which produce aliphatic acids and gases (Ward and Zeikus, 1980; Hillis, 1987).

With increasing radial growth of the trunk, commensurate increases in crown size occur, resulting in the enlargement of existing branches and the production of new ones; crown development is not only outwards but upwards. Radial

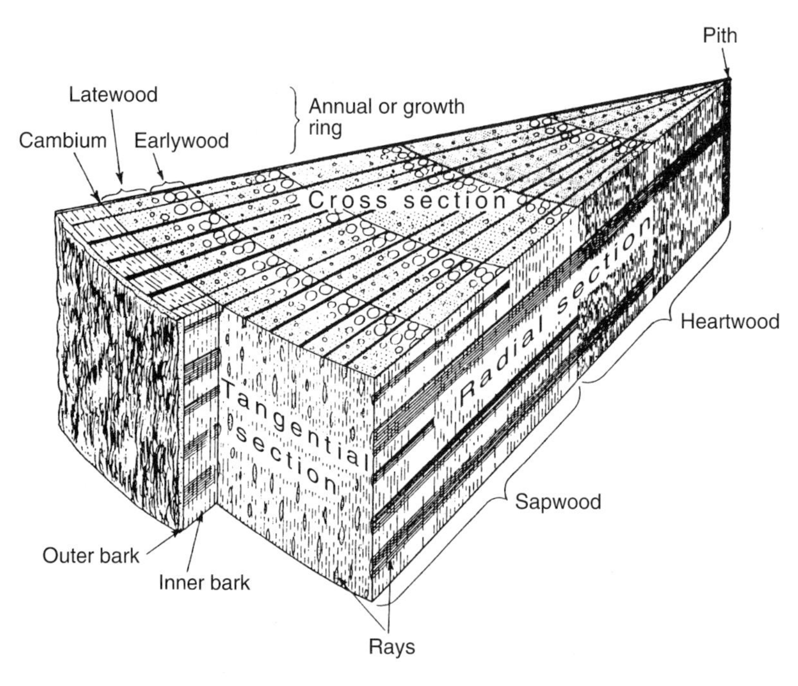

Figure 1.1 Diagramatic illustration of a wedge-shaped segment cut from a 5-year-old hardwood tree, showing the principal structural features. (© BRE.)

growth of the trunk must accommodate the existing branches and this is achieved by the structure that we know as the *knot*. If the cambium of the branch is still alive at the point where it fuses with the cambium of the trunk, continuity in growth will arise even though there will be a change in orientation of the cells. The structure so formed is termed a *green* or *live* knot (Figure 1.3). If, however, the cambium of the branch is dead, and this frequently happens to the lower branches, there will be an absence of continuity, and the trunk will grow round the dead branch, often complete with its bark. Such a knot is termed a *black* or *dead* knot (Figure 1.4), and will frequently drop out of planks on sawing. The grain direction in the vicinity of knots is frequently distorted and in a later section the loss of strength due to different types of knots will be discussed.

1.2.2 Structure at the microscopic level

The cellular structure of wood is illustrated in Figures. 1.5 and 1.6. These three-dimensional blocks are produced from micrographs of samples of wood 0.8 × 0.5 × 0.5 mm in size removed from a coniferous tree (Figure 1.5), known technically as a *softwood*, and a broadleaved tree (Figure 1.6), a *hardwood*. In the softwoods about 90% of the cells are aligned in the vertical axis, whereas in

Figure 1.2 Cross-section through the trunk of a Douglas fir tree showing the annual growth rings, the darker heartwood, the lighter sapwood and the bark. (© BRE.)

the hardwoods there is a much wider range in the percentage of cells that are vertical (80–95%). The remaining percentage is present in bands, known as *rays*, aligned in one of the two horizontal planes known as the radial plane or quartersawn plane (Figure 1.1). This means that there is a different distribution of cells on the three principal axes and this is one of the two principal reasons for the high degree of anisotropy present in timber.

It is popularly believed that the cells of wood are living cells. This is certainly not the case. Wood cells are produced by division of the *cambium*, a zone of living cells which lies between the bark and the woody part of the trunk and branches (Figure 1.1). In the winter the cambial cells are dormant and generally consist of a single circumferential layer. With the onset of growth in the spring, the cells in this single layer subdivide radially to form a cambial zone

Figure 1.3 Green or live knot showing continuity in structure between the branch and tree trunk. (© BRE.)

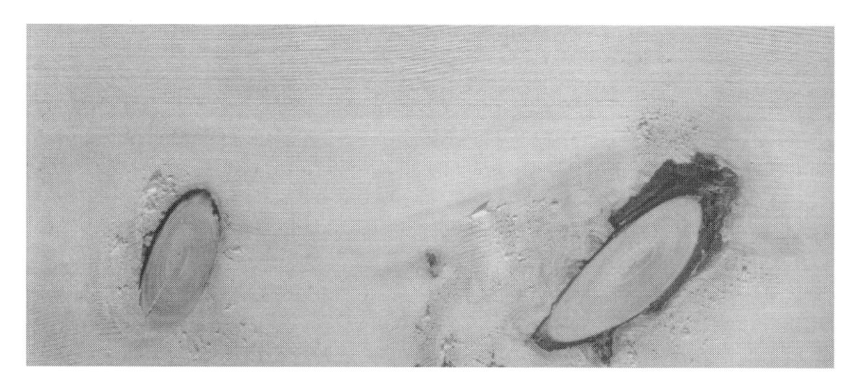

Figure 1.4 Black or dead knot surrounded by the bark of the branch and hence showing discontinuity between branch and tree trunk. (© BRE.)

some 10 cells in width. This is achieved by the formation within each dividing cell of a new vertical wall called the primary wall. During the growing season these cells undergo further radial subdivision to produce what are known as daughter cells and some of these will remain as cambial cells while others, to the outside of the zone, will develop into bark or, if on the inside, will change

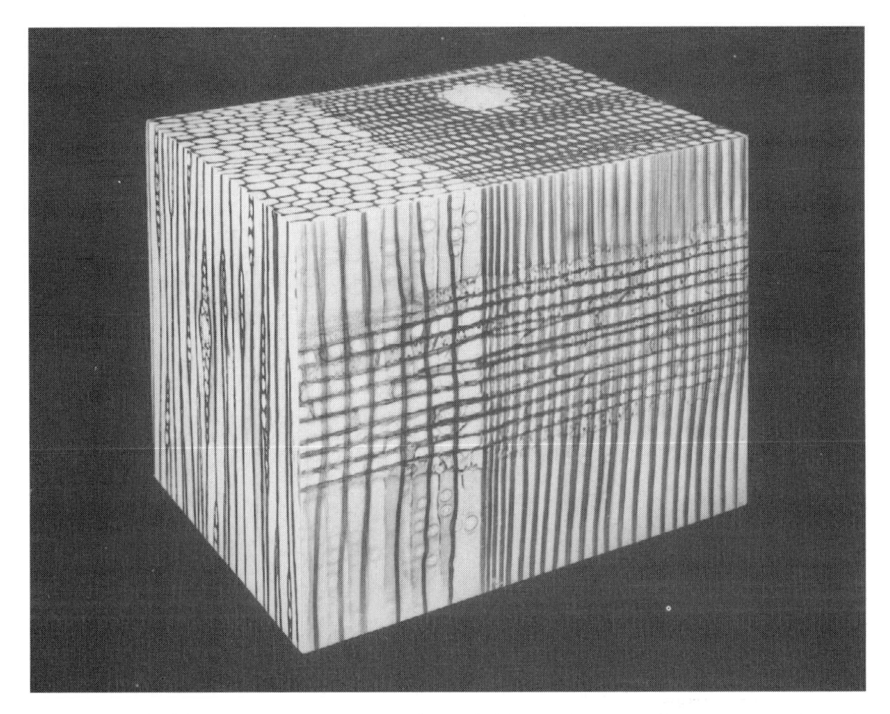

Figure 1.5 Cellular arrangement in a softwood (*Pinus sylvestris* – Scots pine, redwood). (© BRE.)

into wood. There is thus a constant state of flux within the cambial zone with the production of new cells and the relegation of existing cambial cells to bark or wood. Towards the end of the growing season the emphasis is on relegation and a single layer of cambial cells is left for the winter period.

To accommodate the increasing diameter of the tree, the cambial zone must increase circumferentially and this is achieved by the periodic tangential division of the cambial cells. In this case, the new wall is sloping and subsequent elongation of each half of the cell results in cell overlap, often frequently at shallow angles to the vertical axis, giving rise to the formation of spiral grain in the timber. The rate at which the cambium divides tangentially has a significant effect on the average cell length of the timber produced (Bannan, 1954).

The daughter cells produced radially from the cambium undergo a series of changes extending over a period of about three weeks, a process known as *differentiation*. Changes in cell shape are paralleled with the formation of the secondary wall, the final stages of which are associated with the death of the cell; the degenerated cell contents are frequently to be found lining the cell cavity. It is during the process of differentiation that the standard daughter cell is transformed into one of four basic cell types (Table 1.1).

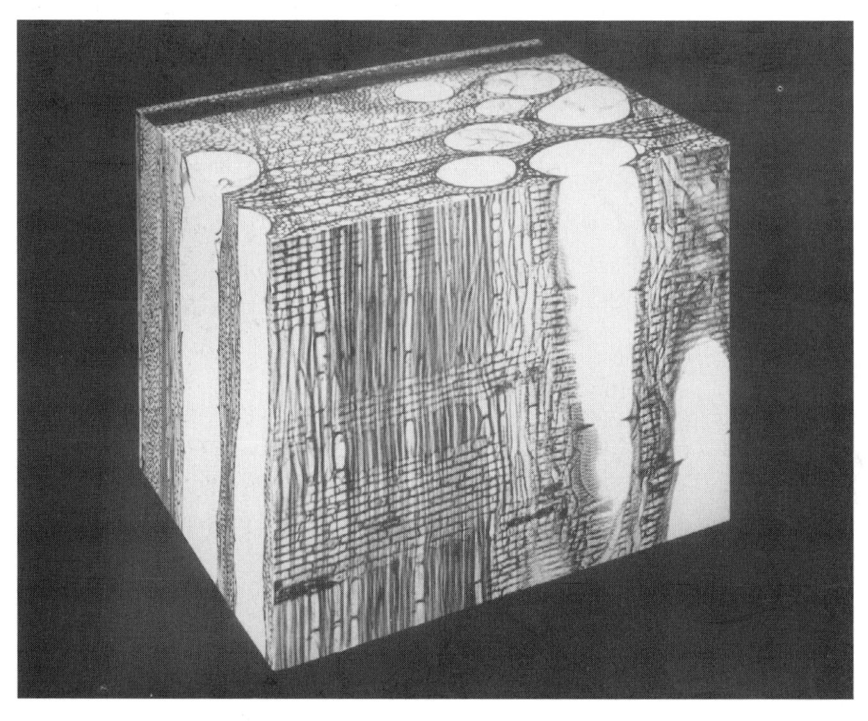

Figure 1.6 Cellular arrangement in a ring-porous hardwood (*Quercus robur* – European oak). (© BRE.)

Table 1.1 Functions and wall thicknesses of the various types of cell found in softwoods and hardwoods

Cells	Softwood	Hardwood	Function	Wall thickness
Parenchyma	✓	✓	storage	
Tracheids	✓	✓	support, conduction	
Fibres		✓	support	
Vessels (pores)		✓	conduction	

Chemical dissolution of the lignin–pectin complex cementing the cells together will result in their separation and this is a useful technique for separating and examining individual cells. In the softwood (Figure 1.7) two types of cell can be observed. Those present in greater number are known as *tracheids*, some 2–4 mm in length with an aspect ratio (L/D) of about 100:1. These cells, which lie vertically in the tree trunk, are responsible for both the supporting and conducting roles. The small block-like cells some 200 × 30 μm in size, known as *parenchyma*, are mostly located in the *rays* and are responsible for the storage of food material.

In contrast, in the hardwoods (Figure 1.8), four types of cell are present albeit that one, the tracheid, is present in small amounts. The role of storage is again primarily taken by the parenchyma, which can be present horizontally in the form of a ray, or vertically, either scattered or in distinct zones. Support is effected by long thin cells with very tapered ends, known as *fibres*; these are usually about 1–2 mm in length with an aspect ratio of about 100:1. Conduction is carried out in cells whose end walls have been dissolved away either completely or in part. These cells, known as *vessels* or *pores*, are usually short

Figure 1.7 Individual softwood cells (× 20). (© BRE.)

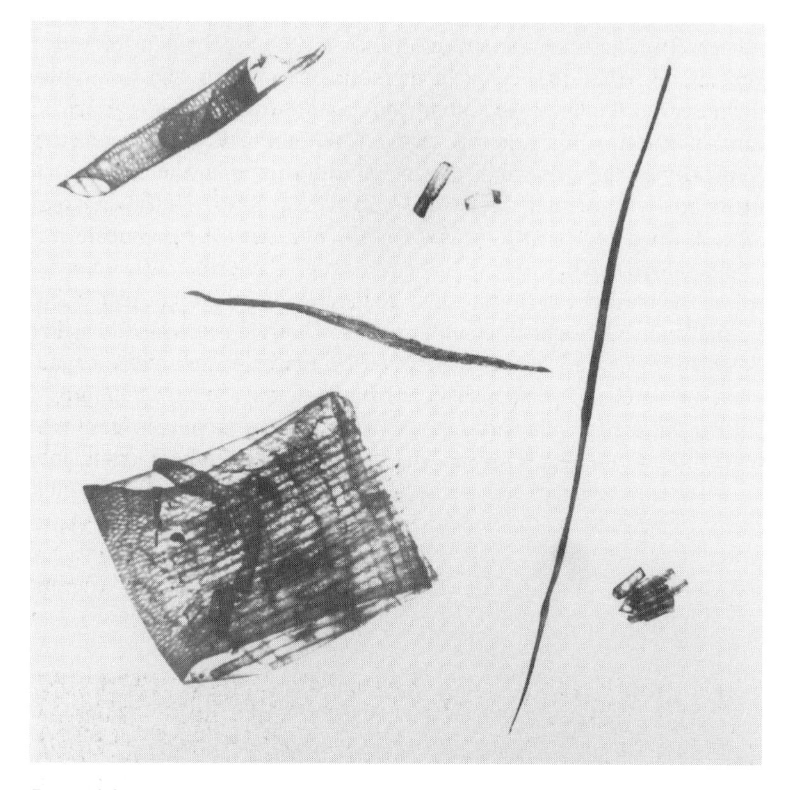

Figure 1.8 Individual cells from a ring-porous hardwood (× 50). (© BRE.)

(0.2–1.2 mm) and relatively wide (up to 0.5 mm) and when situated above one another form an efficient conducting tube. It can be seen, therefore, that whereas in the softwoods the three functions are performed by two types of cell, in the hardwoods each function is performed by a single cell type (Table 1.1).

Although all cell types develop a secondary wall this varies in thickness, being related to the function that the cell will perform. Thus, the wall thickness of fibres is several times that of the vessel (Table 1.1). Consequently, the density of the wood, and hence many of the strength properties discussed later, are related to the relative proportions of the various types of cell. Density, of course, will also be related to the absolute wall thickness of any one type of cell, for it is possible to obtain fibres of one species of wood with a cell wall several times thicker than those of another. The range in density of timber is from 120 to 1200 kg/m^3, corresponding to pore volumes of from 92% to 18% (see Chapter 3).

Growth may be continuous throughout the year in certain parts of the world and the wood formed tends to be uniform in structure. In the temperate and subarctic regions and in parts of the tropics growth is seasonal, resulting in the formation of *growth rings*; where there is a single growth period each year these rings are referred to as *annual rings* (Figure 1.1).

When seasonal growth commences, the dominant function appears to be conduction, whereas in the latter part of the year the dominant factor is support. This change in emphasis manifests itself in the softwoods with the presence of thin-walled tracheids (about 2 μm) in the early part of the season (the wood being known as *earlywood*) and thick-walled (up to 10 μm) and slightly longer (10%) in the latter part of the season (the *latewood*) (Figure 1.1).

In some of the hardwoods, but certainly not all of them, the earlywood is characterised by the presence of large-diameter vessels surrounded primarily by parenchyma and tracheids; only a few fibres are present. In the latewood, the vessel diameter is considerably smaller (about 20%) and the bulk of the tissue comprises fibres. It is not surprising to find, therefore, that the technical properties of the earlywood and latewood are quite different from one another. Timbers with this characteristic two-phase system are referred to as having a *ring-porous* structure (Figure 1.6).

The majority of hardwoods, whether of temperate or tropical origin, show little differentiation into earlywood and latewood. Uniformity across the growth ring occurs not only in cell size, but also in the distribution of the different types of cells (Figure 1.9); these timbers are said to be *diffuse porous*.

Figure 1.9 Cellular arrangement in a diffuse-porous hardwood (*Fagus sylvatica* – beech). (© BRE.)

In addition to determining many of the technical properties of wood, the distribution of cell types and their sizes is used as a means of timber identification. Interconnection by means of pits occurs between cells to permit the passage of mineral solutions and food in both longitudinal and horizontal planes. Three basic types of pit occur. *Simple pits*, generally small in diameter and taking the form of straight-sided holes with a transverse membrane, occur between parenchyma and parenchyma, and also between fibre and fibre. Between tracheids a complex structure known as the *bordered pit* occurs (Figure 1.10; see also Figure 5.4(a) for sectional view). The entrance to the pit is domed and the internal chamber is characterised by the presence of a diaphragm (the *torus*) which is suspended by thin strands (the *margo*). Differential pressure between adjacent tracheids will cause the torus to move against the pit aperture, effectively stopping flow. As discussed later, these pits have a profound influence on the degree of artificial preservation of the timber. Similar structures are to be found interconnecting vessels in a horizontal plane. Between parenchyma cells and tracheids or vessels, *semi-bordered* pits occur and are often referred to as ray pits. These are characterised by the presence of a dome on the tracheid or vessel wall and the absence of such on the parenchyma wall; a pit membrane is present, but the torus is absent. Differences in the shape and size of these pits is an important diagnostic feature in the softwoods.

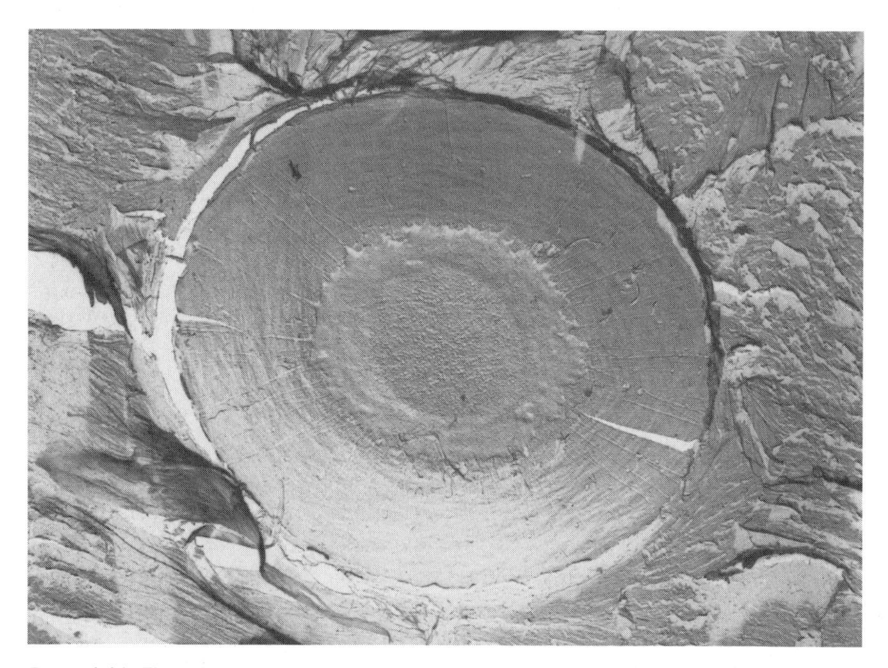

Figure 1.10 Electron micrograph of the softwood bordered pit showing the margo strands supporting the diaphragm (torus), which overlaps the aperture (× 3600). (© BRE.)

The general arrangement of the vertically aligned cells is referred to as *grain*. Although it is often convenient when describing timber at a general level to regard these cells as lying truly vertical, this is not really true in the majority of cases; these cells generally deviate from the vertical axis in a number of different patterns.

In many timbers, and certainly in most of the softwoods, the direction of the deviation from the vertical axis is consistent and the cells assume a distinct spiral mode which may be either left- or right-handed (Harris, 1989). In young trees the helix is usually left-handed and the maximum angle which is near to the core is frequently of the order of 4°, though considerable variability occurs both within a species and also between different species of timber. As the trees grow, so the helix angle in the outer rings decreases to zero and quite frequently in very large trees the angle in the outer rings subsequently increases, but the spiral has changed hand. Spiral grain has very significant technical implications; the strength is lowered, and the degree of twisting on drying and the amount of pick-up on machining increase as the degree of spirality of the grain increases (Brazier, 1965).

In other timbers the grain can deviate from the vertical axis in a number of more complex ways of which *interlocked* and *wavy* are perhaps the two most common and best known. As each of these types of grain deviation give rise to a characteristic decorative figure, further discussion on grain is reserved until the next chapter.

1.2.3 Molecular structure and ultrastructure

1.2.3.1 Chemical constituents

One of the principal tools used to determine the chemical structure of materials is X-ray diffraction analysis. In this technique a beam of X-rays is pinpointed onto the substance and the diffracted beam is recorded on a photosensitive emulsion. The nature, position and intensity of the image of this beam are indications of the degree, type and orientation of the crystallinity present in the material. An analysis of this type on a small block of timber produces an image similar to that illustrated in Figure 1.11. The diffuse zone in the centre of the image is characteristic of an amorphous or non-crystalline material, whereas the band of higher intensity indicates the presence of some crystalline material. The fact that the band is segmented rather than continuous shows that this crystalline material is orientated in a particular plane, the angle of orientation being determined by measurement of the intensity distribution on either the paratropic planes (002), (101) and $(10\bar{1})$, or the diatropic plane (040). These techniques have been developed and refined over the years and now embrace elaborate computer software in the analysis and modelling of the data (see e.g. Meylan, 1967; El-Osta *et al.*, 1973; Paakkari and Serimaa, 1984; Cave 1997). Other techniques that have been used in analysing the fine structure of the cell

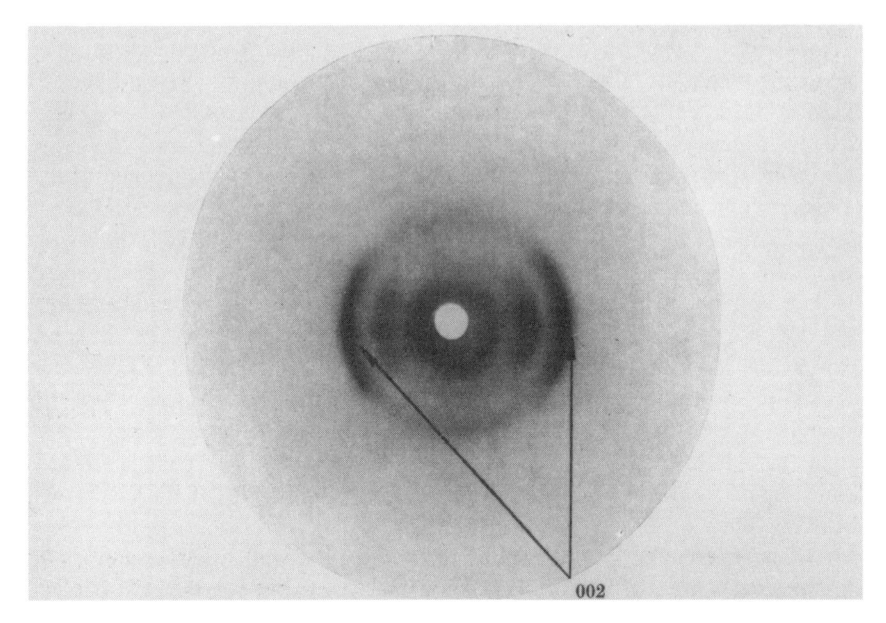

Figure 1.11 X-ray diagram of wood showing diffuse central halo due to the amorphous regions, and the lateral arcs resulting from the crystalline regions. A densitometric trace across the 002 ring provides information from which the microfibrillar angle can be determined. (From D. R. Cowdrey and R. D. Preston (1966) *Proceedings of the Royal Society*, **B166**, 245–272, reproduced by permission of the Royal Society.)

wall are polarisation and electron microscopies, various types of spectroscopy and nuclear magnetic resonance (NMR).

Chemical analysis reveals the existence of four constituents and provides data on their relative proportions. The information revealed by X-ray diffraction and chemical analyses may be summarised as in Table 1.2. The proportions are for timber in general and slight variations in these can occur between timber of different species.

CELLULOSE

Cellulose $(C_6H_{10}O_5)_n$ occurs in the form of long slender filaments or chains, these having been built up within the cell wall from the glucose monomer $(C_6H_{12}O_6)$. Although the number of units per cellulose molecule (the degree of polymerisation) can vary considerably even within one cell wall, it is thought that a value of 8000–10 000 is a realistic average for the secondary cell wall (Goring and Timell, 1962) whereas the primary cell wall has a degree of polymerisation of only 2000–4000 (Simson and Timell, 1978). The anhydroglucose unit $C_6H_{10}O_5$, which is not quite flat, is in the form of a six-sided ring consisting

Table 1.2 Chemical composition of timber

Component	*Mass*		*Polymeric state*	*Molecular derivatives*	*Function*
	Softwood (%)	*Hardwood (%)*			
Cellulose	42±2	45±2	crystalline, highly oriented, large linear molecule	glucose	'fibre'
Hemicelluloses	27±2	30±5	semicrystalline, smaller molecule	galactose mannose xylose	'matrix'
Lignin	28±3	20±4	amorphous, large 3-D molecule	phenylpropane	'matrix'
Extractives	3±2	5±4	principally compounds soluble in organic solvents	terpenes, polyphenols, stilbenoids	extraneous

of five carbon atoms and one oxygen atom (Figure 1.12). The side groups play an important part in intra- and intermolecular bonding as described later. Successive glucose units are covalently linked in the 1,4 positions giving rise to a potentially straight and extended chain; moving in a clockwise direction around the ring it is the first and fourth carbon atoms after the oxygen atom that combine with adjacent glucose units to form the long-chain molecule. The anhydroglucose units comprising the molecule are not flat; rather, they assume a chair configuration with the hydroxyl groups (one primary and two secondary) in the equatorial positions and the hydrogen atoms in the axial positions (Figure 1.12).

Glucose can be present in one of two forms dependent on the position of the –OH group attached to carbon 1. When this lies above the ring, i.e. on the same side as that on carbon 4, the unit is called α-glucose and when this combines with an adjacent unit with the removal of H–O–H (known as a condensation reaction) the resulting molecule is called starch, a product manufactured in the crown and stored in the parenchyma cells.

When the –OH group lies below the ring, the unit is known as β-glucose and on combining with adjacent units, again by a condensation reaction, a molecule of cellulose is produced. It is this product that is the principal wall-building constituent of timber. Usually the linkage of β-units is stronger than that of α-units but, in order to achieve the former, alternate anhydroglucose units must be rotated through 180°.

Cellulose chains may crystallise in many ways, but one form, namely cellulose I, is characteristic of natural cellulosic materials. In early studies on cellulosic materials including wood, X-ray diffraction analyses have been interpreted as indicating that the cellulose crystal is characterised by a repeat distance of 1.03 nm, equivalent to two anhydroglucose units, in the chain direction (*b*-axis),

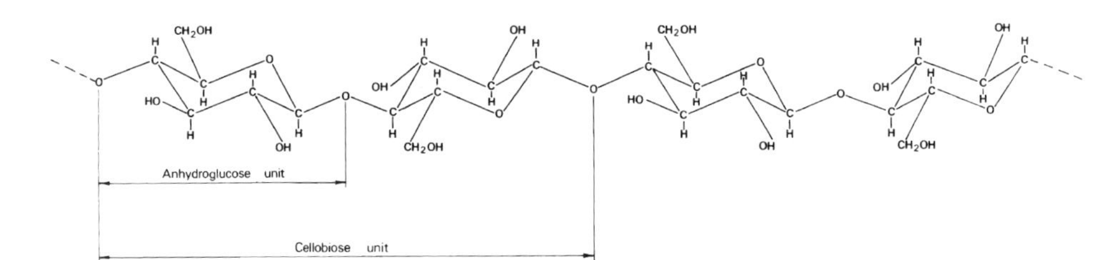

Figure 1.12 Structural formula for the cellulose molecule in its chair configuration. (© BRE.)

with the other edges of the unit cell having lengths of 0.835 nm in the [100] crystallographic direction (*a*-axis) and 0.790 nm in the [001] direction (*c*-axis). The *a*- and *c*- axes are inclined at 84° to each other and both are perpendicular to the *b*-axis (i.e. the crystal is monoclinic).

These spacings and angle have been fitted more or less satisfactorily into a number of unit cells over the years. The one that found most acceptance in the early days was that proposed by Meyer and Misch (1937) incorporating two cellulose chains (Figure 1.13). Four cellobiose units (each comprising two glucose residues) form the edges of the monoclinic unit cell whereas one cellobiose unit passes through the centre; the unit cell, therefore, comprises one whole chain plus four quarter chains. All chains are aligned parallel with the *b*-axis of the unit cell, but with the central chain antiparallel to its neighbours; the plane of the anhydroglucose units lies in the *ab* plane of the unit cell. Within this plane the chains lie close together giving rise along the *a*-axis to strong interchain hydrogen bonding; between these ab planes (i.e. in the direction of the *c*-axis) bonding is limited to only van der Waals forces.

The advent of stereochemistry and the use of space-fitting models were adopted by Hermans *et al.* (1943) and led to the replacement of straight chains in the Meyer and Misch model by chains which were bent along their length both within the plane of the glucose rings and perpendicular to it. This gives rise to intrachain hydrogen bonding (Figure 1.14).

Although there was no direct evidence from studies on timber to indicate that adjacent chains lie in opposite directions, there was considerable indirect evidence to suggest this arrangement. The presence of an antiparallel arrangement has been widely adopted in the past, although since the late 1970s many workers have come to dispute this hypothesis and to indicate that the chains could all lie in the same direction.

Whether the chains are parallel or antiparallel, steric considerations require that the chain at the centre of the unit cell is displaced longitudinally by 0.29 nm, a distance equivalent to *b*/4; this facilitates interchain bonding. An antiparallel displacement does not infer that the cellulose molecule is folded, as is characteristic of most of the man-made crystalline polymers. Although a folded-chain

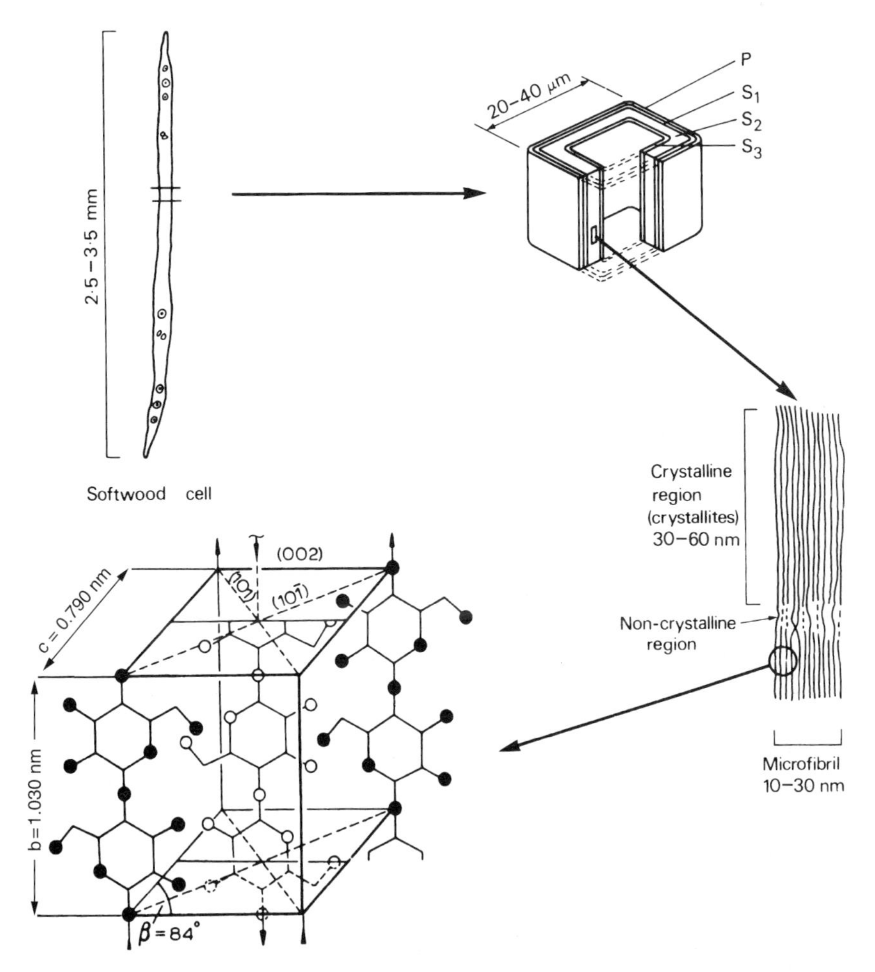

Figure 1.13 The relationship between the structure of wood at different levels of magnitude. The unit cell is the classical model of cellulose I proposed by Meyer and Misch (1937). (Layout adapted from J.F. Siau (1971) *Flow in wood*, reproduced by permission of Syracuse University Press.)

model was proposed for cellulose some time ago the results of more recent work have refuted this hypothesis.

Investigations on the structure of cellulose carried out on the alga *Valonia* in the 1970s have indicated that in this simple plant there is a very high probability that the cellulose chains are all facing the same way (Gardner and Blackwell, 1974; Sarko and Muggli, 1974). Whereas the former workers interpret the unit cell still as monoclinic though with dimensions twice that of the Meyer and Misch cell, thereby incorporating eight chains, the latter workers propose a triclinic cell with new dimensions, but comprising only two chains.

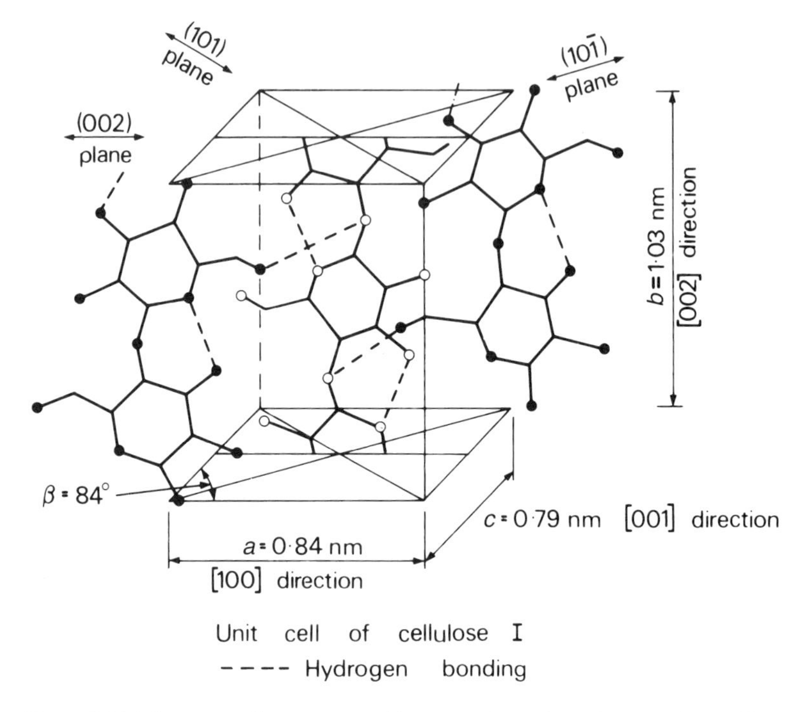

Figure 1.14 The unit cell as proposed by a number of workers in the 1940s to 1960s incorporating antiparallel chains, chains with bent configurations as proposed by Hermans *et al.* (1943), intramolecular bonding, again as proposed by Hermans *et al.*, and intermolecular bonding on the 101 and 10$\bar{1}$ planes as proposed among others by Liang and Marchessault (1959) and Jaswon *et al.* (1968).

The model proposed by Gardner and Blackwell appears to have gained wider acceptance and it is interesting to note that an eight-chain unit cell was proposed many years ago by Honjo and Watanabe (1958) and by Frei and Preston (1961). In using X-ray fibre diffraction methods, Gardner and Blackwell refined the structure using rigid-body least-square methods of analysis to obtain the best agreement between observed and calculated X-ray intensities. These results indicate that in native cellulose (cellulose I) all the chains have the same polarity, i.e. they are parallel, unlike the Meyer and Misch model where the central chain was antiparallel. Forty-one reflections were observed which were indexed using a monoclinic unit cell having dimensions a = 1.634 nm, b =1.572 nm and c = 1.038 nm (the cell axis) with β = 97°. The unit cell therefore comprises a number of whole chains or parts of chains totalling eight in number. (Note that the use of b and c to describe respectively one horizontal axis and the vertical axis by Gardner and Blackwell is transposed from that in the Meyer and Misch model: β is now the complimentary angle.)

All but three of the reflections can be indexed by a two-chain unit cell almost identical to that of Meyer and Misch. These three reflections are reported as being very weak, which means that the differences between the four Meyer and Misch unit cells making up the eight-chain cell must be small. Gardner and Blackwell therefore take a two-chain unit cell (a = 0.817 nm, b = 0.786 nm and c = 1.038 nm) as an adequate approximation to the real structure. Their proposed model for cellulose I is shown in Figure 1.15, which shows the chains lying in a parallel configuration; the centre chain is staggered by $0.266 \times c$ (= 0.276 nm).

A parallel configuration continues to be widely though not unanimously accepted; it is interesting to note in passing that it was first proposed by Preston as early as 1959. Cellulose that has been regenerated from a solution displays a different crystalline structure and is known as cellulose II. In this case there is complete agreement that the unit cell possesses an antiparallel arrangement of the cellulose molecule.

Within the structure of cellulose I, both primary and secondary bonding is represented and many of the technical properties of wood can be related to the variety of bonding present. Covalent bonding both within the glucose rings and linking together the rings to form the molecular chain contributes to the high axial tensile strength of timber. There is no evidence of primary bonding laterally between the chains; rather, this seems to be a complex mixture of the fairly strong hydrogen bonds and the weak van der Waals forces. The same OH groups that give rise to this hydrogen bonding are highly attractive to water molecules and explain the affinity of cellulose for water. Whereas Meyer and Misch (1937) had placed the hydrogen bonds within the (002) *ab* plane, it was later thought (now incorrectly) by a number of workers that these bonds united the cellulose chains of the (101) and ($10\bar{1}$) planes (Jaswon *et al.*, 1968; Figure 1.14), thereby providing the main mechanism for stabilising the crystal against relative displacement of the chain and consequently contributing considerably to the axial stiffness of wood. Gardner and Blackwell (1974) on cellulose from *Valonia* identify the existence of both intermolecular and intramolecular hydrogen bonds all of which, however, are interpreted as lying only on the (020) plane (Figure 1.15); they consider the structure of cellulose as an array of hydrogen-bonded sheets held together by van der Waals forces and this concept is widely, though not universely, held at present.

The degree of crystallinity of the cellulose is usually assessed by X-ray and electron diffraction techniques though other methods have been employed. Generally, a value of about 60% is obtained, although values as high as 90% are recorded in the literature. This wide range in values is due in part to the different techniques employed in the determination of crystallinity and in part to the fact that wood is comprised not just of the crystalline and noncrystalline constituents, but rather a series of substances of varying crystallinity. Regions of complete crystallinity and regions with a total absence of crystalline structure (amorphous zones) can be recognised, but the transition from one state to the other is gradual.

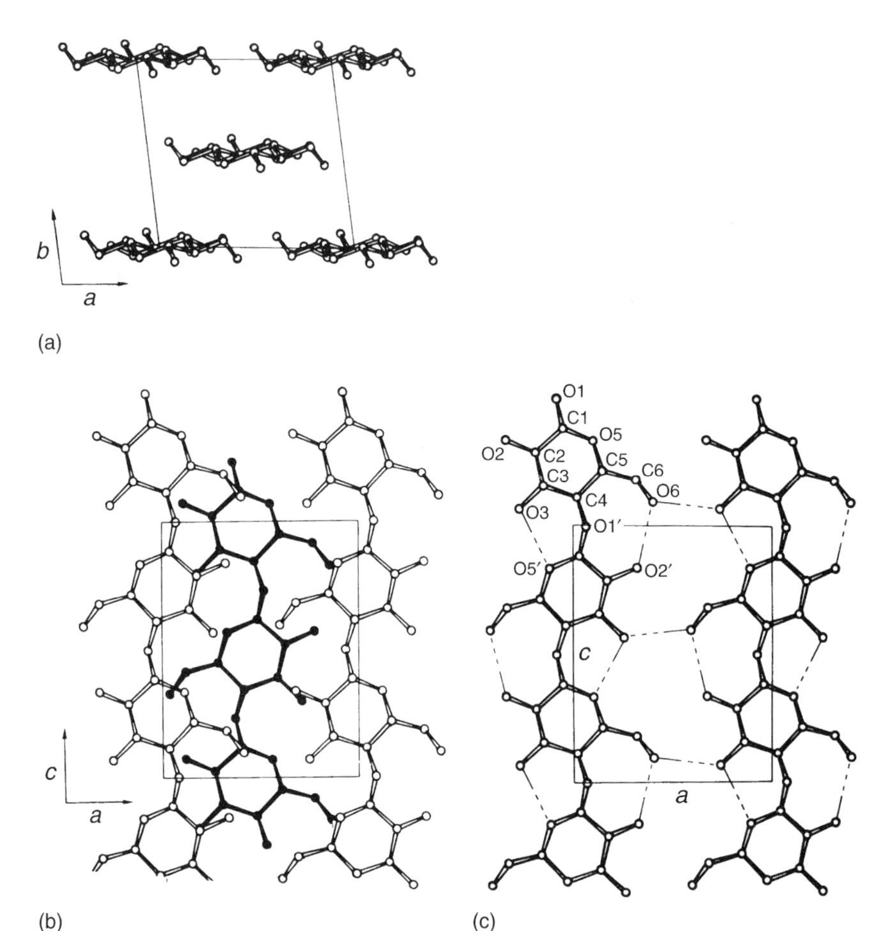

Figure 1.15 Projections of the proposed parallel two-chain model for cellulose I. (a) Unit cell viewed perpendicular to the *ab* plane (along the fibre axis). (b) Unit cell viewed perpendicular to the *ac* plane. Note that the central chain (in black) has the same orientation as the other chains and is staggered vertically with respect to them by an amount equal to *c*/4. (c) Projection of the (002) plane, only showing the hydrogen bonding network. Each glucose residue forms two intramolecular bonds (O3–H · · · O5′), and (O2′–OH · · · O6), and one intermolecular bond (O6–H · · · O3). (From Gardner and Blackwell (1974), *Biopolymers* 13,1975–2001, reproduced by permission of John Wiley and Sons, Inc.) Note that the use of *b* and *c* to describe respectively one horizontal axis and the vertical axis by Gardner and Blackwell is transposed from that in the Meyer and Misch model (Figure 1.13) and that β is now the complementary angle.

The length of the cellulose molecule is about 5000 nm (0.005 mm) whereas the average size of each crystalline region determined by X-ray analysis is only 60 nm in length, 5 nm in width and 3 nm in thickness. This means that any cellulose molecule will pass through several regions of high crystallinity – known as *crystallites* or *micelles* – with intermediate noncrystalline or low-crystalline zones in which the cellulose chains are in only loose association with each other (Figure 1.13). Thus, the majority of chains emerging from one crystallite will pass to the next, creating a high degree of longitudinal coordination; this collective unit is termed a *microfibril* and has infinite length. It is clothed with chains of cellulose mixed with chains of sugar units other than glucose (see below) which lie parallel, but are not regularly spaced. This brings the microfibril in timber to about 10 nm in breadth, and in some algae, such as *Valonia*, to 30 nm. The degree of crystallinity will therefore vary along its length and it has been proposed that this could be periodic.

Readers desirous of a more comprehensive account of the historical investigation of the structure of cellulose are referred to the review by Hon (1994).

HEMICELLULOSES AND LIGNIN

In Table 1.2 reference was made to the other constituents of wood additional to cellulose. Two of these, the hemicelluloses and lignin, are regarded as cementing materials contributing to the structural integrity of wood and also to its high stiffness. The hemicelluloses, like cellulose itself, are carbohydrates built up of sugar units, but are unlike cellulose in the type of units they comprise. These units differ between softwoods and hardwoods, and generally the total percentage of the hemicelluloses present in timber is greater in hardwoods compared with softwoods (Table 1.2). Thus, in the softwoods the hemicellulose fraction comprises (by mass of the timber) from 15% to 20% of galactoglucomannans and about 10% of arabinoglucoronoxylan. By contrast, the hardwoods contain 20–30% of glucoronoxylan and 5% glucomannan. Both the degree of crystallisation and the degree of polymerisation of the hemicelluloses are generally low, the molecule containing less than 200 units; in these respects, and also in their lack of resistance to alkali solutions, the hemicelluloses are quite different from true cellulose (Siau, 1984).

Lignin, present in about equal proportions to the hemicelluloses, is chemically dissimilar to these and to cellulose. Lignin is a complex, three-dimensional, aromatic molecule composed of phenyl groups with a molecular weight of about 11 000. It is non-crystalline and the structure varies between wood from a conifer and from a broadleaved tree. Softwood lignin is composed mostly of guaiacyl-type lignin whereas the hardwood lignin comprises both guaiacyl and syringyl units. Over the years there has been great controversy on whether or not these units are present as a copolymer, or whether they are present separately and, if the latter, whether the two components are found in different tissues of the timber. There is evidence (e.g. Obst 1982) to suggest that the ratio of syringyl to guaiacyl lignin does not differ in different cells and that they are probably present as a copolymer.

About 25% of the total lignin in timber is to be found in the middle lamella, an intercellular layer composed of lignin and pectin, together with the primary cell wall. As this compound middle lamella is very thin, the concentration of lignin is correspondingly high (about 70%). Deposition of the lignin in this layer is rapid.

The bulk of the lignin (about 75%) is present within the secondary cell wall, having been deposited following completion of the cellulosic framework. Initiation of lignification of the secondary wall commences when the compound middle lamella is about half completed and extends gradually inwards across the secondary wall (Saka and Thomas, 1982). Termination of the lignification process towards the end of the period of differentiation coincides with the death of the cell. Most cellulosic plants do not contain lignin and it is the inclusion of this substance within the framework of timber that is largely responsible for the stiffness of timber, especially in the dried condition.

EXTRACTIVES

Before leaving the chemical composition of wood, mention must be made of the presence of extractives (Table 1.2). This is a collective name for a series of highly complex organic compounds present in certain timbers in relatively small amounts. Some, such as waxes, fats and sugars, have little economic significance, but others, for example rubber and resin (from which turpentine is distilled), are of considerable economic importance. The heartwood of timber, as described previously, generally contains extractives which, in addition to imparting colouration to the wood, bestow on it its natural durability, as most of these compounds are toxic to both fungi and insects. Readers desirous of more information on extractives are referred to the excellent text by Hillis (1987).

MINERALS

The presence of minerals in timber also contributes to its chemical composition. Elements such as calcium, sodium, potassium, phosphorus and magnesium are all components of new growth tissue, but the actual mass of these inorganic materials is small and constitutes on the basis of the oven-dry mass of the timber less than 1% for temperate woods and less than 5% for tropical timbers.

Certain timbers show a propensity to conduct suspensions of minerals that are subsequently deposited within the timber. The presence of silica in the rays of certain tropical timbers (see Figure 9.2, for example), and calcium carbonate in the cell cavities of iroko are two examples where large concentrations of minerals cause severe problems in log conversion and subsequent machining.

ACIDITY

Wood is generally acidic in nature, the level of acidity being considerably higher in the heartwood compared to the sapwood of the same tree. The pH of the heartwood varies in different species of timber, but is generally about 4.5–5.5; however, in some timbers such as eucalypt, oak, and western red cedar, the pH of the heartwood can be as low as 3.0. Sapwood generally has a pH at least 1.0 higher than the corresponding heartwood, i.e. the acidity is at least 10 times lower than the corresponding heartwood.

Acidity in wood is due primarly to the generation of acetic acid by hydrolysis of the acetyl groups of the hemicelloses in the presence of moisture. These acetyl groups can be present in levels corresponding to between 2% and 5% of the mass of oven dry wood. The higher acidity of the heartwood may be due in some small part to the presence of certain extractives, but is much more likely to reflect the higher concentrations of acetic acid. Acidity in wood can cause severe corrosion of certain metals and care has to be exercised in the selection of metallic fixings; the level of corrosion is much greater at higher relative humidities.

Small quantities of formic acid are also formed, but its effect can be ignored in comparison with that of acetic acid.

1.2.3.2 The cell wall as a fibre composite

In the introductory remarks, wood was defined as a natural composite and the most successful model used to interpret the ultrastructure of wood from the various chemical and X-ray analyses ascribes the role of 'fibre' to the cellulosic microfibrils, while the lignin and hemicelluloses are considered as separate components of the 'matrix'. The cellulosic microfibril, therefore, is interpreted as conferring high tensile strength to the composite owing to the presence of covalent bonding both within and between the anhydroglucose units. Experimentally it has been shown that reduction in chain length following gamma irradiation markedly reduces the tensile strength of timber (Ifju, 1964); the significance of chain length in determining strength has been confirmed in studies of wood with inherently low degrees of polymerisation. Although Ifju considered slippage between the cellulose chains to be an important contributor to the development of ultimate tensile strength, this is thought to be unlikely due to the forces involved in fracturing large numbers of hydrogen bonds.

Preston (1964) has shown that the hemicelluloses are usually intimately associated with the cellulose, effectively binding the microfibrils together. Bundles of cellulose chains are therefore seen as having a polycrystalline sheath of hemicellulose material and consequently the resulting high degree of hydrogen bonding would make chain slippage unlikely; rather, it would appear that stressing results in fracture of the C–O–C linkage.

The deposition of lignin is variable in different parts of the cell wall, but it is obvious that its prime function is to protect the hydrophilic (water-seeking) non-crystalline cellulose and the hemicelluloses which are mechanically weak when wet. Experimentally, it has been demonstrated that removal of the lignin markedly reduces the strength of wood in the wet state, though its reduction results in an increase in its strength in the dry state calculated on a net cell wall area basis. Consequently, the lignin is regarded as lying to the outside of the microfibril, forming a protective sheath.

As the lignin is located only on the exterior it must be responsible for cementing together the fibrils and in imparting shear resistance in the transference of stress throughout the composite. The role of lignin in contributing towards stiffness of timber has already been mentioned.

There has been great debate over the years as to juxtaposition of the cellulose, hemicellulose and lignin in the composition of a microfibril, and to the size of the basic unit. Two of the many models proposed are illustrated in Figure 1.16. The model in Figure 16(a) depicts cellulosic subunits some 3 nm in diameter. These units, comprising some 40 cellulose chains, are known as elementary fibrils or protofibrils. Gaps (1 nm) between these units are filled with hemicellulose while more hemicellulose and lignin form the sheath.

In passing, it is interesting to note that subunits as small as 1 nm (subelementary fibril) have been claimed by some researchers (Hanna and Côté, 1974). Subdivision of the microfibril in this way was frequently disputed in the early days as it was thought that the evidence to support such a subdivision had been produced by artefacts in sample preparation for electron microscopy.

The model in Figure 16(b) shows the crystalline core to be about 5 nm × 3 nm containing about 48 chains in either four- or eight-chain unit cells; this latter configuration has received wider acceptance. Both these models, however, are in some agreement in that passing outwards from the core of the microfibril the highly crystalline cellulose gives way first to the partly crystalline layer containing mainly hemicellulose and non-crystalline cellulose, and then to the amorphous lignin. This gradual transition of crystallinity from fibre to matrix results in high interlaminar shear strength which contributes considerably to the high tensile strength and toughness of wood.

1.2.3.3 Cell wall layers

When a cambial cell divides to form two daughter cells a new wall is formed comprising the middle lamella and two primary cell walls, one to each daughter cell. These new cells undergo changes within about 3 days of their formation and one of these developments will be the formation of a secondary wall. The thickness of this wall will depend on the function the cell will perform, as described earlier, but its basic construction will be similar in all cells.

Early studies on the anatomy of the cell wall used polarisation microscopy, which revealed the direction of orientation of the crystalline regions (Preston, 1934). These studies indicated that the secondary wall could be subdivided into three layers and measurements of the extinction position was indicative of the

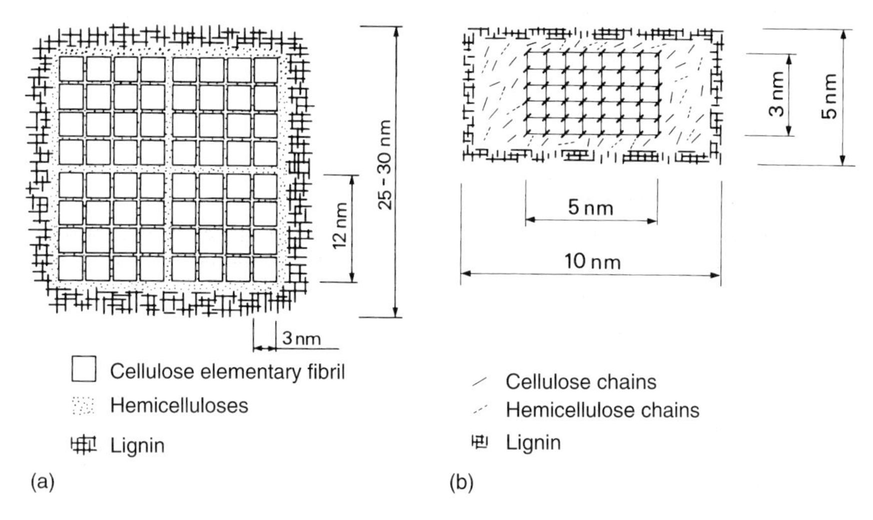

Figure 1.16 Models of the cross-section of a microfibril: (a) the crystalline core has been subdivided into elementary fibrils, whereas in (b) the core is regarded as being homogeneous. ((a) adapted from D. Fengel (1970) *The Physics and Chemistry of Wood Pulp Fibres 1970*, TAPPI, with permission; (b) adapted from R.D. Preston (1974) *The Physical Biology of Plant Cell Walls*, reproduced by permission of Chapman and Hall.)

angle at which the microfibrils were orientated. Subsequent studies with transmission electron microscopy confirmed these findings and provided some additional information with particular reference to wall texture and variability of angle. However, much of our knowledge on microfibrillar orientation has been derived using X-ray diffraction analysis, on either the paratropic (002) plane or the diatropic (040) plane. Most of these techniques yield only mean angles for any one layer of the cell wall, but recent analysis has indicated that it may be possible to determine the complete microfibrillar angle distribution of the cell wall (Cave, 1997).

The relative thickness and mean microfibrillar angle of the layers in a sample of spruce timber are illustrated in Table 1.3.

Table 1.3 Microfibrillar orientation and percentage thickness of the cell wall layers in spruce timber (*Picea abies*)

Wall layer	*Approximate thickness (%)*	*Angle to longitudinal axis*
P	3	random
S_1	10	50–70°
S_2	85	10–30°
S_3	2	60–90°

The middle lamella, a lignin–pectin complex, is devoid of cellulosic microfibrils whereas in the primary wall (P) the microfibrils are loose packed and interweave at random (Figure 1.17); no lamellation is present. In the secondary wall layers the microfibrils are closely packed and parallel to each other. The outer layer of the secondary wall, the S_1 is again thin and is characterised by having from four to six concentric lamellae, the microfibrils of each alternating between a left- and right-hand spiral (*S* and *Z* helix) both with a pitch to the longitudinal axis of from 50° to 70° depending on the species of timber.

The middle layer of the secondary wall (S_2) is thick and is composed of 30–150 lamellae, the closely packed microfibrils of which all exhibit a similar orientation in a right-hand spiral (*Z* helix) with a pitch of 10°–30° to the longitudinal axis, as illustrated in Figure 1.18. As over three-quarters of the cell wall

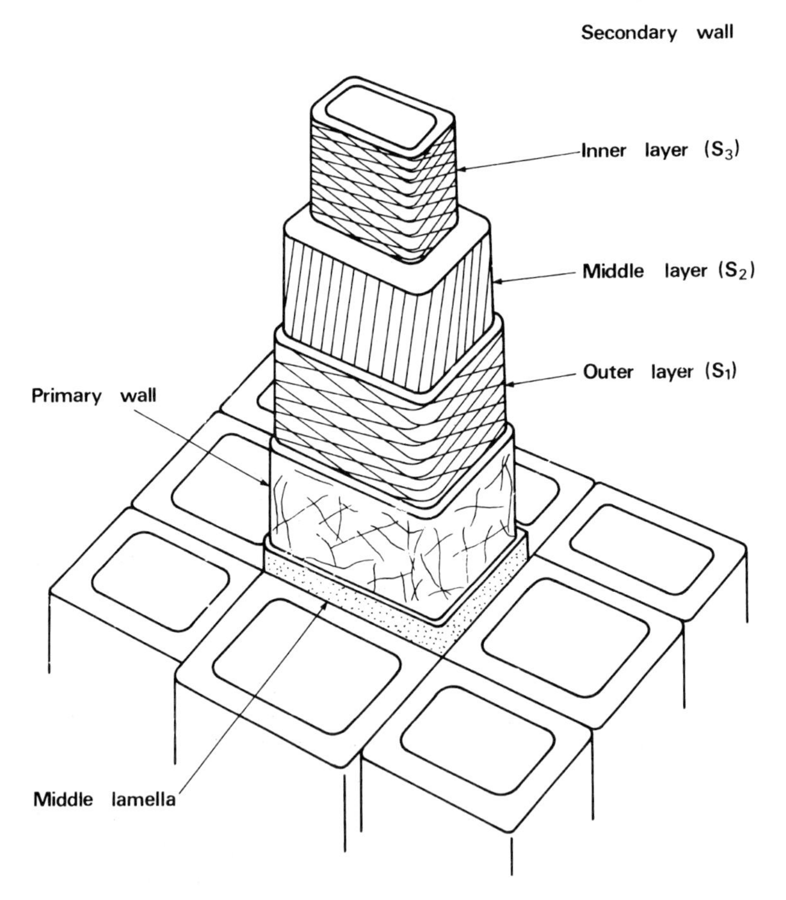

Figure 1.17 Simplified structure of the cell wall showing mean orientation of microfibrils in each of the major wall layers. (© BRE.)

Figure 1.18 Electron micrograph of the cell wall in Norway spruce timber (*Picea abies*) showing the parallel and almost vertical microfibrils of an exposed portion of the S_2 layer. (© BRE)

is composed of the S_2 layer, it follows that the ultrastructure of this layer will have a very marked influence on the behaviour of the timber. In later sections, anisotropic behaviour, shrinkage, tensile strength and failure morphology will all be related to the microfibrillar angle in the S_2 layer.

Kerr and Goring (1975) were among the first workers to question the extent of these concentric lamellae in the S_2 layer. Using electron microscopy to examine sections of permanganate-stained spruce tracheids, these workers found that though there was a preferred orientation of lignin and carbohydrates in the S_2 layer, the lamellae were certainly not continuous. Thus, the interrupted lamellae model proposed by them embraced lignin and carbohydrate entities which were greater in the tangential than in the radial direction. Cellulose microfibrils were envisaged as being embedded in a matrix of hemicelluloses and lignin.

The existence within the S_2 layer of concentric lammellae has been questioned again in the last few years. Evidence has been presented (Sell and Zinnermann, 1993; Sell, 1994) from both electron and light microscopy which indicates radial, or near radial orientations of the transverse structure of the S_2 layer. The transverse thickness of these agglomerations of microfibrils is 0.1–1.0 nm and they frequently extend the entire width of the S_2 layer. A modified model of the cell wall of softwoods has been proposed (Sell and Zinnerman, 1993).

The S_3 layer, which may be absent in certain timbers, is very thin with only a few concentric lamellae. It is characterised, as in the S_1 layer, by alternate lamellae possessing microfibrils orientated in opposite spirals with a pitch of 60–90°, though the presence of the right-handed spiral (*Z* helix) is disputed by some workers. Generally, the S_3 has a looser texture than the S_1 and S_2 layers, and is frequently encrusted with extraneous material. The S_3, like the S_1, has a higher concentration of lignin than in the S_2 (Saka and Thomas, 1982). Electron microscopy has also revealed the presence of a thin warty layer overlaying the S_3 layer in certain timbers.

Investigations have indicated that the values of microfibrillar angle quoted in Table 1.3 are only average for the layers and that systematic variation in angle occurs within each layer. The inner lamellae of the S_1 tend to have a smaller angle, and the outer lamellae a larger angle than the average for each layer. A similar but opposite situation occurs in the S_3 layers. Variation also occurs between different species and systematically within a species; thus, the microfibrillar angle is usually greater in the earlywood compared with the corresponding latewood, and significantly greater in the longitudinal–radial cell wall than in the longitudinal–tangential wall.

Since the late 1970s and with the advent of much-improved electron microscopy, many investigators have recorded the presence of two very thin transition layers, an S_{12} layer between the S_1 and S_2 layers and an S_{23} between the S_2 and S_3 layers; the microfibrillar angle of these transition layers is intermediate between that of the adjacent wall layers. It is most unlikely that these very thin layers contribute to the behaviour of timber and it is therefore convenient to continue to treat the cell wall as comprising only three principal layers.

In a recent investigation on *Abies sachalinensis* using field emission electron microscopy (Abe *et al*., 1991) not only was the existence of systematic variation across each wall layer confirmed, but the magnitude and nature of this variation was also shown to be considerable. Rather than each lamella of the S_1 layer having microfibrils in both the left- and right-hand spirals (*S* and *Z* helices), as was previously thought to be the case, this investigation demonstrated that the outermost microfibrils of this layer were in a left-hand (*S*) spiral of 45°. This angle increased systematically inwards to reach 90° to the longitudinal cell axis before the orientation of the spiral was reversed to a right-hand one (*Z*). Further inward movement saw this right-hand spiral decrease to 70°. The S_2 layer was confirmed as having closely packed microfibrils in a right-hand spiral with an angle of between 0° and 20°, whereas the S_3 layer was shown to be almost a mirror image of the newly described S_1. The outer layers had microfibrils in a right-hand spiral of 70° which increased in successive lamellae to 90° to the longitudinal cell axis before decreasing in a left-hand spiral to a minimum of 30° on the inner boundary of the wall. The microfibrillar angle of the secondary wall, as seen from the lumen, therefore changed in a clockwise direction from the outermost S_1 to the middle of the S_2 and there in a counter clockwise direction to the innermost S_3. This resulted in the boundaries between the three principal layers being very indistinct, confirming reports by previous workers on other species.

The microfibrillar angle appears to vary systematically along the length of the cell as well as across the wall thickness. Thus, the angle of the S_2 layer has been shown to decrease towards the ends of the cells, whereas the average S_2 angle appears to be related to the length of the cell (Preston,1934), itself a function of rate of growth of the tree. Systematic differences in the microfibrillar angle have been found between the radial and tangential walls, and this has been related to differences in the degree of lignification between these walls. Openings occur in the walls of cells and many of these pit openings are characterised by localised deformations of the microfibrillar structure.

Over the years a number of hypotheses have been put forward to explain the differences in microfibril orientation within plant cell walls. Of these, only the multinet growth hypothesis, which postulates that the reorientation of microfibrils is due to cell extension and which was originally proposed by Roelofsen and Houwink (1953), has a substantial number of advocates. Many scientists, however, are sceptical of its validity, because it is incompatible with various microfibril arrangements. Boyd (1985a), from a very extensive examination and reappraisal of existing data and views, has prepared a lengthy and well-argued monograph in which he concludes that cell wall development involves biophysical factors that necessarily prevent multinet's postulated large reorientation of microfibrils after their formation. A new theory is formulated based on the strains generated in the cell wall and plasmalemma as a consequence of extension growth and subsequent wall thickening. For wood this new strain theory is shown to support explanations for differences in microfibrillar orientation between different wall layers.

Booker and Sell (1998) have adopted a different approach in evaluating reasons for the particular microfibrillar angles in the S_1, S_2 and S_3 layers. They have elucidated these with reference to the mechanical requirements for the successful functioning of trees; for example, a physical relationship has been shown to exist between trans-wall cracking and microfibrillar angle.

1.2.4 *Variability in structure*

Variability in performance of wood is one of its inherent deficiencies as a material. Chapters 6 and 7 discuss how differences in mechanical properties occur between timbers of different species and how these are manifestations of differences in wall thickness and distribution of cell types. However, superimposed on this genetical source of variation is both a systematic and an environmental one.

There are distinct patterns of variation in many features within a single tree. The length of the cells, the thickness of the cell wall and hence density, the angle at which the cells are lying with respect to the vertical axis (spiral grain), and the angle at which the microfibrils of the S_2 layer of the cell wall are located with respect to the vertical axis, all show systematic trends outwards from the centre of the tree to the bark and upwards from the base to the top of the tree. This pattern results in the formation of a core of wood in the tree with many undesirable properties including low strength and high shrinkage. This zone, usually regarded as some 10–20 growth rings in width, is known as *core* wood or *juvenile* wood as opposed to the *mature* wood occurring outside this area. The boundary between juvenile and mature wood is usually defined in terms of the change in slope of the variation in magnitude of one anatomical feature (such as cell length or density) when plotted against ring number from the pith. In a recent study comprising 180 Douglas fir trees, the period of juvenile wood when defined in terms of density variation was found to vary by as much as 11–37 rings from the pith. It was concluded from the analysis that the magnitude of the juvenile period, at least as far as these trees were concerned, was under appreciable genetic control (Abel-Gadir and Krahmer, 1993).

In another study on loblolly pine, the demarcation between juvenile and mature wood was determined using segmented modelling, iterative solution, and constrained solution techniques in the analysis of specific gravity data. This indicated that the boundary usually occurred at about 11 rings from the pith; the value was consistent with height within any one tree, but did vary with site (Tasissa and Burkhart, 1998).

Within the living trunk of both softwood and hardwood trees, internal stresses are generated as a result of microscopic dimensional changes in the wood structure. Two theories have been put forward to explain the origin of these *growth stresses*. The first theory explains their origin in terms of longitudinal contraction of the cells as they swell laterally due to the deposition of lignin in the interstices of the wall (Boyd 1985b), whereas the second theory states that longitudinal tensile stress in the cells is generated naturally from the contraction of cellulosic microfibrils due probably to crystalisation of the cellulose and that it is subsequently reduced or modified by the deposition of lignin (Bamber 1979 and 1987). In large diameter trees and especially in certain genera, the magnitude of these stresses can generate major problems in log conversion and subsequent machining of the timber. These growth stresses display systematic patterns of variation – in the longitudinal radial plane the core region is in longitudinal compression, whereas the outer layers are in tension. In the transverse radial plane the stresses are in tension, whereas in the transverse tangential plane they are in compression. Figure 1.19 illustrates these patterns and provides guidance on the level of stresses that may occur in large diameter eucalypts.

Mention has already been made of the systematic variation in many timbers in grain angle outward from the pith (Section 1.2.2).

Environmental factors have considerable influence on the structure of the wood and any environmental influence, including forest management, which changes the rate of growth of the tree will affect the technical properties of the wood. However, the relationship is a complex one; in softwoods, increasing growth rate generally results in an increase in the width of earlywood with a resulting decrease in density and mechanical properties. In diffuse-porous hardwoods increasing growth rate, provided it is not excessive, has little effect on density, whereas in ring-porous hardwoods, increasing rate of growth, again provided it is not excessive, results in an increase in the width of latewood and consequently in density and strength.

There is a whole series of factors which may cause defects in the structure of wood and consequent lowering of its strength. Perhaps the most important defect with regard to its utilisation is the formation of *reaction wood*. When trees are inclined to the vertical axis, usually as a result of wind action or of growing on sloping ground, the distribution of growth-promoting hormones

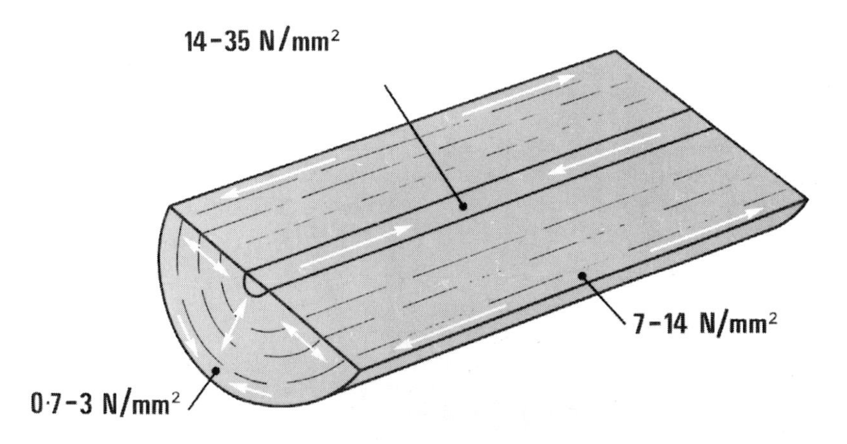

Figure 1.19 The range of growth stresses and direction of action in logs of *Eucalyptus* spp.(© BRE.)

is disturbed, resulting in the formation of an abnormal type of tissue. In the softwoods, this reaction tissue grows on the compression side of the trunk and is characterised by having a higher than normal lignin content, a higher microfibrillar angle in the S_2 layer resulting in increased longitudinal shrinkage, and a generally darker appearance (Figure 1.20). This abnormal timber, known as *compression wood*, is also considerably more brittle than normal wood. In the hardwoods, reaction wood forms on the tension side of trunks and large branches and is therefore called *tension wood*. It is characterised by the presence of a gelatinous cellulosic layer (the *G layer*) to the inside of the cell wall; this results in a higher than normal cellulose content to the total cell wall which imparts a rubbery characteristic to the fibres resulting in difficulties in sawing and machining (Figure 1.21).

One other defect of considerable technical significance is *brittleheart*, which is found in many low-density tropical hardwoods. Due to the slight shrinkage of cells after their formation, the outside layers of the tree are in a state of longitudinal tension resulting in the cumulative increase of compression stress in the core. A time is reached in the growth of the tree when the compression stresses due to growth (one of the growth stresses described earlier) are greater than the natural compression strength of the wood. Yield occurs with the formation of shear lines through the cell wall and throughout the core wood. Compression failure is discussed in greater detail in a later section.

Readers desirous of further information on the important area of variability in structure and its influence on the technical performance of timber are referred to Chapters 5 and 12 of Desch and Dinwoodie (1996).

Figure 1.20 A band of compression wood (below centre) in a Norway spruce plank, illustrating the darker appearance and higher longitudinal shrinkage of the reaction wood compared with the adjacent normal wood. (© BRE.)

Figure 1.21 Board of African mahogany showing rough surface and concentric zones of well developed tension wood (the lighter zones) on the end grain. (© BRE.)

References

Anon (1998) A reference for the forestry industry. 1997 *Yearbook of the Forestry Industry Council of Great Britain*, Stirling.

Abe, H., Ohtani, J. and Fukazawa, K. (1991) FE-SEM observations on the microfibrillar orientation in the secondary wall of tracheid. *IAWA Bull. New series*, **12**(4), 431–438.

Abel-Gadir, A.Y. and Krahmer, R.L. (1993). Genetic variation in the age of demarcation between juvenile and mature wood in Douglas-fir. *Wood and Fiber Sci.*, **25**(4), 384–394.

Bamber, R.K.(1979) The origin of growth stresses. *Forpride Digest*, **8**, 75–79, 96.

Bamber, R.K.(1987) The origin of growth stresses; a rebuttal. *IAWA Bull. New series*, **8**(1), 80–84.

Bannan, M.W. (1954) Ring width, tracheid size and ray volume in stem wood of *Thuja occidentalis*. *Can. J. Bot.*, **32**, 466–479.

Booker, R.E. and Sell, J. (1998) The nanostructure of the cell wall of softwoods and its functions in the living tree. *Holz als Roh-und Werkstoff*, **56**,1–8.

Boyd, J.D. (1985a) *Biophysical control of microfibril orientation in plant cell walls. Aquatic and terrestrial plants including trees*, Martinus Nijhoff/Dr W. Junk Publishers, Dordrecht, The Netherlands.

Boyd, J.D. (1985b) The key factor in growth stress generation in trees. Lignification or crystalisation? *IAWA Bull. New series*, **6**, 139–150.

Brazier, J. (1965) An assessment of the incidence and significance of spiral grain in young conifer trees. *For. Prod. J.*, **15**(8) 308–312.

Cave, I.D. (1997) Theory of X-ray measurement of microfibril angle in wood. Part 2 The diffraction diagram, X-ray diffraction by materials with fibre type symmetry, *Wood Sci. Technol.*, **31**, 225–234.

Cowdrey, D.R. and Preston, R.D. (1966) Elasticity and microfibrillar angle in the wood of Sitka spruce, *Proc. Roy. Soc.*, **B166**, 245–272.

Desch, H.E. and Dinwoodie, J.M. (1996) *Timber – structure, properties, conversion and use*, 7th edn, Macmillan, Basingstoke.

El-Osta, M.L., Kellogg, R.M., Foschi, R.O. and Butters, R.G. (1973) A direct X-ray method for measuring microfibril angle, *Wood and Fiber*, **5** (2), 118–128.

Fengel, D. (1970) The ultrastructural behaviour of cell polysaccharides, in *The Physics and Chemistry of Wood Pulp Fibers*, TAPPI, STAP 8, 74–96.

Frei, E. and Preston, R.D. (1961) Cell wall organisation and wall growth in the filamentous green algae *Cladophora* and *Chaetomorpha*: Part 1, The basic structure and its formation. *Proc. Roy. Soc.*, **B154**, 70–94; Part 2, Spiral structure and spiral growth. *Proc. Roy. Soc.*, **B155**, 55–77.

Gardner, K.H. and Blackwell, J. (1974) The structure of native cellulose. *Biopolymers*, **13**, 1975–2001.

Goring, D.A.I. and Timmell, T.E. (1962) Molecular weight of native cellulose. *TAPPI*, **45**, 454–459.

Hanna R.B. and Côté, W.A. jnr (1974) The sub-elementary fibril of plant cell wall cellulose. *Cytobiologie*, **10** (1), 102–116.

Harris, J.M. (1989) *Spiral grain and wave phenomena in wood formation*, Springer-Verlag, Berlin.

Hermans, P.H., De Booys, J. and Maan, C.H. (1943) Form and mobility of cellulose molecules. *Kolloid Zeitschrift*, **102**, 169–180.

Hillis, W.E. (1987) *Heartwood and tree exudates*, Springer-Verlag, Berlin.

Hon, D.N-S (1994) Cellulose; a random walk along its historical path. *Cellulose*, **1**, 1–25.

Honjo, G. and Watanabe, M. (1958) Examination of cellulose fiber by the low temperature specimen method of electron diffraction and electron microscopy. *Nature*, London, **181** (4605) 326–328.

Ifju, G. (1964) Tensile strength behaviour as a function of cellulose in wood, *For. Prod. J.*, **14**, 366–372.

Jaswon, M.A., Gillis, P.P. and Marks, R.E. (1968) The elastic constants of crystalline native cellulose. *Proc. Roy. Soc.*, **A306**, 389–412.

Kerr, A.J. and Goring, D.A.I. (1975) Ultrastructural arrangement of the wood cell wall. *Cellulose Chem. Techol.*, **9** (6), 563–573.

Liang, J.Y. and Marchessault, R.H. (1959) Hydrogen bonds in native cellulose. *J. Polym. Sci.*, **35**, 529–530.

Meyer, K.H. and Misch, L. (1937) Position des atomes dans le nouveau modele spatial de la cellulose. *Helv. Chim. Acta*, **20**, 232–244.

Meylan, B.A. (1967) Measurement of microfibrillar angle in *Pinus radiata* by X-ray diffraction. *For. Prod. J.*, **19**, 30–34.

Obst, J.R. (1982) Guaiacyl and syringyl lignin composition in hardwood cell components. *Holzforschung*, **36**, 143–152.

Paakkari, T. and Serimaa, R. (1984) A study of the structure of wood cells by X-ray diffraction. *Wood Sci. Technol.*, **18**, 79–85.

Preston, R.D. (1934) The organisation of the cell wall of the conifer tracheid, *Philosophical Transactions*, **B224**, 131–174.

Preston, R.D. (1959) Wall organisation in plant cells. In: *International Review of Cytology*, Vol. 8, pp. 33–60, Academic Press, New York.

Preston, R.D. (1964) Structural and mechanical aspects of plant cell walls, in *The Formation of Wood in Forest Trees*, H.M. Zimmermann (Ed.), Academic Press, New York. pp. 169–188.

Preston, R.D. (1974) *The Physical Biology of Plant Cell Walls*, Chapman & Hall, London.

Roelofsen, P.A. and Houwink, A.L. (1953) Architecture and growth of the primary cell wall in some plant hairs in the *Phycomyces sporangiophore. Acta Bot. Néerl.*, **2**, 218–225.

Saka, S. and Thomas, R.J. (1982) A study of lignification in Loblolly pine tracheids by the SEM-EDXA technique. *Wood Sci. Technol.*, **12**, 51–62.

Sarko, A. and Muggli, R. (1974) Packing analyses of carbohydrates and polysaccharides, III: *Valonia* cellulose and cellulose II. *Macromolecules*, **7** (4), 486–494.

Sell, J. (1994) Confirmation of a sandwich-like model of the cell wall of softwoods by light microscope. *Holz, Roh-Werkstoff*, **57**, 234.

Sell, J and Zinnermann, T. (1993) Radial fibril agglomeration of the S_2 on transverse fracture surfaces of tracheids of tension-loaded spruce and white fir. *Holz, Roh-Werkstoff*, **51**, 384.

Siau, J.F. (1971) *Flow in wood*, Syracuse University Press.

Siau, J. (1984) *Transport processes in* wood, Springer-Verlag, Berlin.

Simson, B.W. and Timell, T.E. (1978) Polysaccharides in cambial tissues of *Populus tremuloides* and *Tilia americana*. V. Cellulose. *Cellul. Chem. Technol.*, **12**, 51–62.

Tasissa, G. and Burkhart, H.E. (1998) Juvenile-mature wood demarcation in loblolly pine trees. *Wood and Fiber Sci.*, **30** (2), 119–127.

Ward J.C. and Zeikus J.G. (1980). Bacteriological, chemical and physical properties of wetwood in living trees. In: Bauch J. (Ed), natural variations of wood properties. *Mitt Bundesforschungstalt Forst-Holzwirtsch* **131**: 133–166 (technical report).

Chapter 2

Appearance of timber in relation to its structure

2.1 Introduction

The previous chapter described the nature or basic structure of timber, whereas this and subsequent chapters are concerned with the behaviour of timber, defined in terms of its properties and performance. Most important of all in these chapters is the presentation of timber behaviour in terms of its nature by adopting a material science approach to the subject.

Most readers will agree that many timbers are aesthetically pleasing and the various and continuing attempts to simulate the appearance of timber in the surface of synthetic materials bear testament to the very attractive appearance of many timbers. Although a very large proportion of the timber consumed in the UK is used within the construction industry, where the natural appearance of timber is of little consequence, excepting the use of hardwoods for flush doors, internal panelling and wood-block floors, a considerable quantity of timber is still utilised purely on account of its attractive appearance particularly for furniture and various sports goods. The decorative appearance of many timbers is due to the *texture*, or to the *figure*, or to the *colour* of the material and, in many instances, to combinations of these.

2.2 Texture

The texture of timber depends on the size of the cells and on their arrangement. A timber such as boxwood in which the cells have a very small diameter is said to be *fine-textured*, whereas a *coarse-textured* timber, such as keruing, has a considerable percentage of large-diameter cells. Where the distribution of the cell-types or sizes across the growth ring is uniform, as in beech, or where the thickness of the cell wall remains fairly constant across the ring, as in some of the softwoods, such as yellow pine, the timber is described as being *even-textured*. Conversely, where variation occurs across the growth ring, either in distribution of cells (as in teak) or in thickness of the cell walls (as in larch or Douglas fir), the timber is said to have an *uneven texture*.

2.3 Figure

Figure is defined as the 'ornamental markings seen on the cut surface of timber, formed by the structural features of the wood' (BS 565 (1972)) but the term is also frequently applied to the effect of marked variations in colour. The four most important structural features inducing figure are *grain*, *growth rings*, *rays* and *knots*.

2.3.1 Grain

Mention was made in the previous chapter that the cells of wood, though often described as vertically orientated, frequently deviate from this convenient concept. In the majority of cases this deviation takes the form of a spiral, the magnitude of the angle varying with distance from the pith. Although of considerable technical importance because of loss in strength and induced machining problems, the common form of spiral grain has no effect on the figure presented on the finished timber. However, two other forms of grain deviation do have a very marked influence on the resulting figure of the wood. Thus, in certain hardwood timbers, and the mahoganies are perhaps the best example, the direction of the spiral in the longitudinal–tangential plane alternates from left to right hand at very frequent intervals along the radial direction; grain of this type is said to be *interlocked*. Tangential faces of machined timber will be normal, but the radial face will be characterised by the presence of alternating light and dark longitudinal bands produced by the reflection of light from the tapered cuts of fibres inclined in different directions (Figure 2.1). This type of figure is referred to as *ribbon* or *stripe* and is desirous in timber for furniture manufacture.

If instead of the grain direction alternating from left to right within successive layers along the radial direction as above, the grain direction alternates at right angles to this pattern, i.e. in the longitudinal–radial plane, a *wavy* type of grain is produced. This is very conspicuous in machined tangential faces where it shows up clearly as alternating light and dark horizontal bands (Figure 2.2); this type of figure is described as *fiddleback*, because timber with this distinctive type of figure has been used traditionally for the manufacture of the backs of violins. It is to be found also on the panels and sides of expensive wardrobes and bookcases.

2.3.2 Growth rings

Where variability occurs across the growth ring, either in the distribution of the various cell types or in the thickness of the cell walls, distinct patterns will appear on the machined faces of the timber. Such patterns, however, will not be regular like many of the man-made imitations, but will vary according to changes in width of the growth ring and in the relative proportions of early- and latewood.

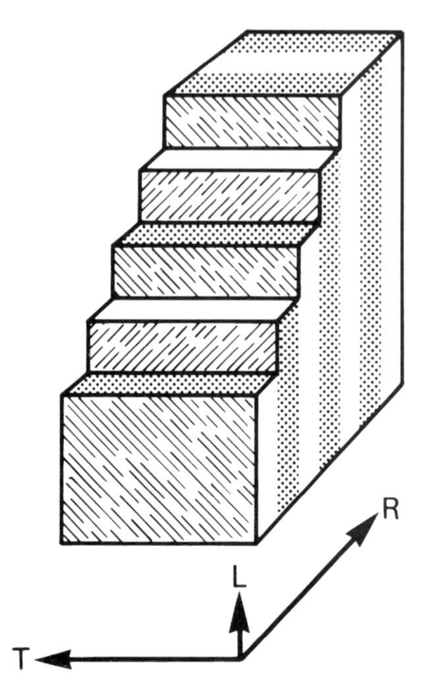

Figure 2.1 Diagramatic illustration of the development of interlocked grain in certain hardwoods. The fibres in successive radial zones are inclined in opposite directions, thereby imparting a striped appearance on the longitudinal–radial plane. (© BRE.)

On the radial face the growth rings will be vertical and parallel to one another, but on the tangential face a most pleasing series of concentric arcs is produced as successive growth layers are intersected. In the centre part of the plank of timber illustrated in Figure 2.3, the growth rings are cut tangentially forming these attractive arcs, while the edge of the board with parallel and vertical growth rings reflects timber cut radially. In the case of elm it is the presence of the large earlywood vessels that makes the growth ring so conspicuous, whereas in timbers such as Douglas fir or pitch pine, the striking effect of the growth ring can be ascribed to the very thick walls of the latewood cells.

2.3.3 Rays

Another structural feature that may add to the attractive appearance of timber is the ray, especially where, as in the case of oak, the rays are both deep and wide. When the surface of the plank coincides with the longitudinal radial plane, these rays can be seen as sinuous light-coloured ribbons running across the grain.

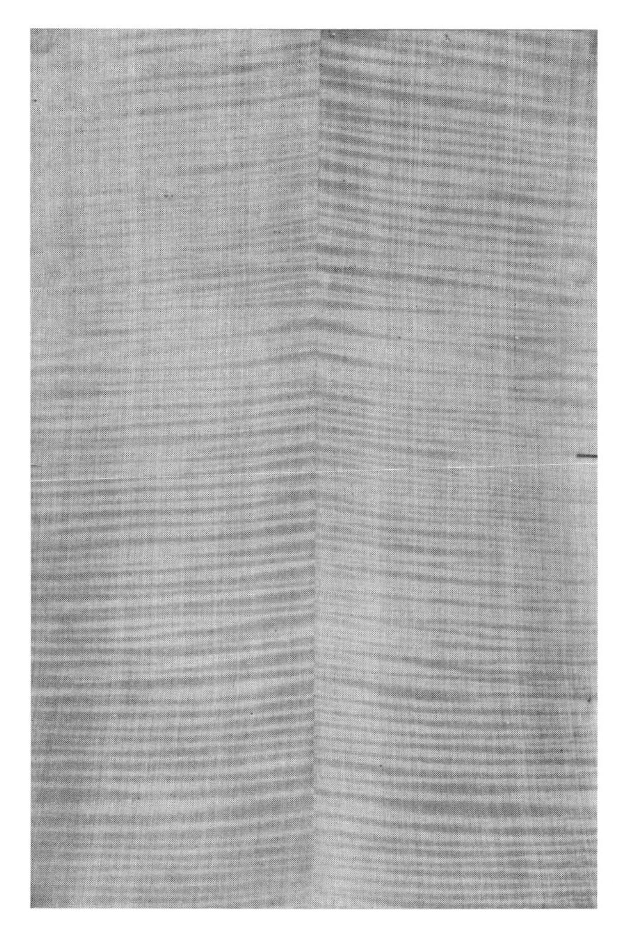

Figure 2.2 'Fiddleback' figure due to wavy grain. (© BRE.)

2.3.4 Knots

Knots, though troublesome from the mechanical aspects of timber utilisation, can be regarded as a decorative feature; the fashion of knotty-pine furniture and wall panelling in the early 1970s is a very good example of how knots can be a decorative feature. However, as a decorative feature, knots do not possess the subtlety of variation in grain and colour that arises from the other structural features described above.

Exceptionally, trees produce a cluster of small shoots at some point on the trunk and the timber subsequently formed in this region contains a multitude of small knots. Timber from these *burrs* is highly prized for decorative work, especially if walnut or yew.

Figure 2.3 The effect of growth rings on figure. (© BRE.)

2.4 Colour

In the absence of extractives, timber tends to be a rather pale straw colour which is characteristic of the sapwood of almost all timbers. The onset of heartwood formation in many timbers is associated with the deposition of extractives, most of which are coloured, thereby imparting colouration to the heartwood zone. In passing, it should be recalled that although a physiological heartwood is always formed in older trees, extractives are not always produced; thus, the heartwood of timbers such as ash and spruce is colourless.

Where colouration of the heartwood occurs, a whole spectrum of colour exists among the different species. The heartwood may be yellow (e.g. boxwood), orange (e.g. opepe), red (e.g. mahogany), purple (e.g. purpleheart), brown (e.g. African walnut), green (e.g. greenheart) or black (e.g. ebony). In some timbers the colour is fairly evenly distributed throughout the heartwood, whereas in other species considerable variation in the intensity of the colour occurs. In zebrano distinct dark brown and white stripes occur, whereas in olive wood patches of yellow merge into zones of brown. Dark gum-veins, as present in African walnut, contribute to the pleasing alternations in colour. Variations in colour such as these are regarded as contributing to the 'figure' of the timber.

It is interesting to note in passing that the non-coloured sapwood is frequently coloured artificially to match the heartwood, thereby adding to the amount of timber converted from the log. In a few rare cases, the presence of certain fungi in timber in the growing tree can result in the formation of very dark coloured heartwood. The activity of the fungus is terminated when the timber is dried.

Both *brown oak* and *green oak*, produced by different fungi, have always b
prized for decorative work.

Reference

BS 565 (1972) *Glossary of terms relating to timber and woodwork*, BSI, London.

erence 41

een

ıme relationships

3.1 Density

The density of a piece of timber is a function not only of the amount of wood substance present, but also the presence of both extractives and moisture. In a few timbers extractives are completely absent, whereas in many they are present, but only in small amounts and usually less than 3% of the dry mass of the timber. In some exceptional cases, the extractive content may be as high as 10% and in these cases it is necessary to remove the extractives prior to the determination of density.

The presence of moisture in timber not only increases the mass of the timber, but it also results in the swelling of the timber, and hence both mass and volume are affected. Thus, in the determination of density where

$$\rho = \frac{m}{v} \tag{3.1}$$

both the mass m and volume v must be determined at the same moisture content. Generally, these two parameters are determined at zero moisture content. However, as density is frequently quoted at a moisture content of 12%, because this level is frequently experienced in timber in use, the value of density at zero moisture content is corrected for 12% if volumetric expansion figures are known, or else the density determination is carried out on timber at 12% moisture content. Thus, if

$$m_x = m_0 \ (1+0.01\mu) \tag{3.2}$$

where m_x is the mass of timber at moisture content x, m_0 is the mass of timber at zero moisture content, and μ is the percentage moisture content, and

$$v_x = v_0 \ (1 + 0.01s_v) \tag{3.3}$$

where v_x is the volume of timber at moisture content x, v_0 is the volume of timber at zero moisture content, and s_v is the percentage volumetric shrinkage/expansion, it is possible to obtain the density of timber at any moisture content in terms of the density at zero moisture content thus:

$$\rho_x = \frac{m_x}{v_x} = \frac{m_0\,(1 + 0.01\mu)}{v_0(1 + 0.01s_v)} = \rho_0\left(\frac{1 + 0.01\mu}{1 + 0.01s_v}\right) \tag{3.4}$$

As a very approximate rule of thumb, the density of timber increases by approximately 0.5% for each 1.0% increase in moisture content up to 30%. Density therefore will increase, slightly and curvilinearly, up to moisture contents of about 30% as both total mass and volume increase; however, at moisture contents above 30%, the density will increase rapidly and curvilinearly with increasing moisture content because (as explained in Chapter 4) the volume remains constant above this value, as the mass increases.

The determination of density by measurement of mass and volume takes a considerable period of time and over the years a number of quicker techniques have been developed for use where large numbers of density determinations are required. These methods range from the assessment of the opacity of a photographic image which has been produced by either light or β-irradiation passing through a thin section of wood, to the use of a mechanical device (the Pilodyn) which fires a spring-loaded bolt into the timber after which the depth of penetration is measured. In all these techniques, however, the method or instrument has to be calibrated against density values obtained by the standard mass–volume technique.

In Section 1.2.2, timber was shown to possess different types of cell which could be characterised by different values of the ratio of cell-wall thickness to total cell diameter. As this ratio can be regarded as an index of density, it follows that density of the timber will be related to the relative proportions of the various types of cells. Density, however, will also reflect the absolute wall thickness of any one type of cell, as it is possible to obtain fibres of one species of timber the cell wall thickness of which can be several times greater than that of fibres of another species.

Density, like many other properties of timber, is extremely variable; it can vary by a factor of 10, ranging from an average value (at 12% moisture content) of 176 kg/m^3 for balsa, to about 1230 kg/m^3 for lignum vitae (Figure 3.1). Balsa, therefore, has a density similar to that of cork, whereas lignum vitae has a density slightly less than half that of concrete or aluminium. The values of density quoted for different timbers are merely average values – each timber will have a range of densities reflecting differences between earlywood and latewood, between the pith and outer rings, and between trees on the same site. Thus, for example, the density of balsa can vary from 40 to 320 kg/m^3.

In certain publications, reference is made to the *weight* of timber, a term widely used in commerce; it should be appreciated that the quoted values are really densities.

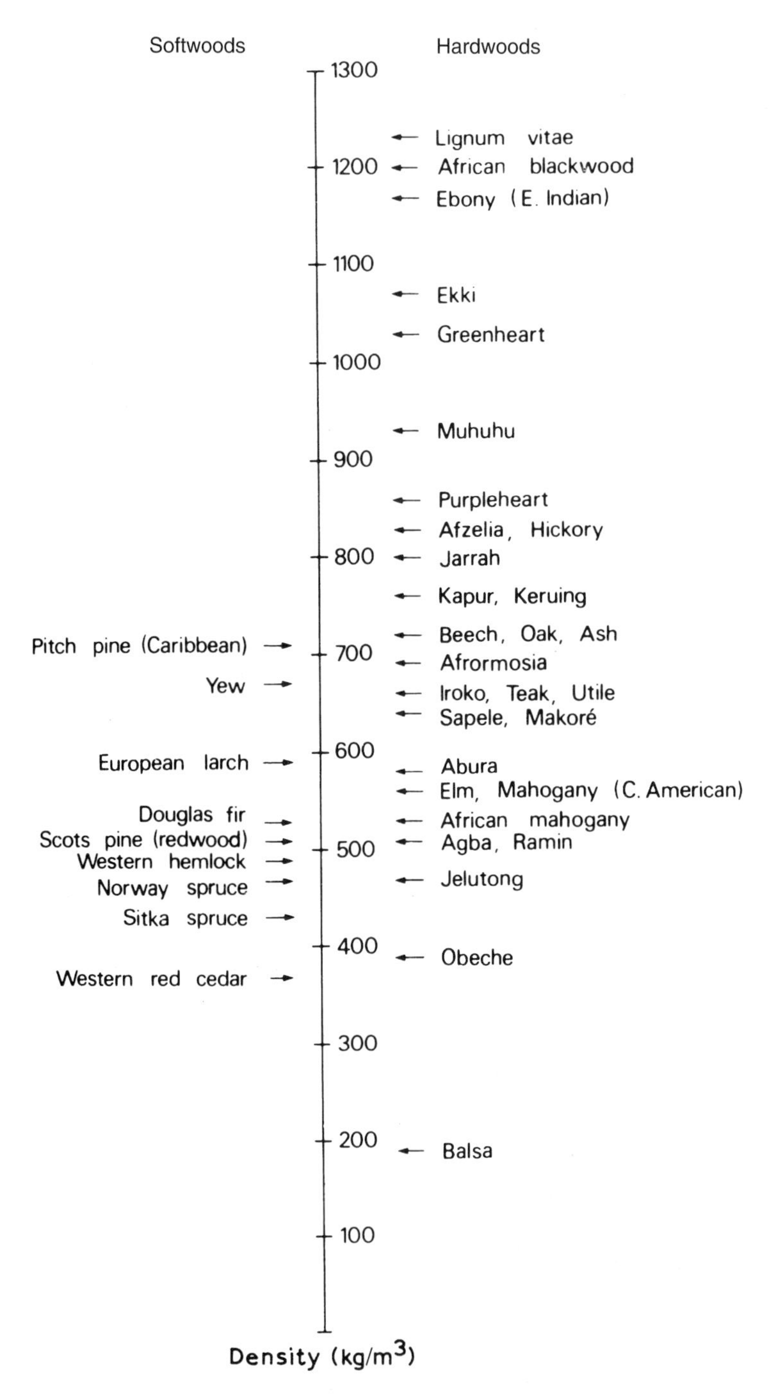

Figure 3.1 Mean density values at 12% moisture content for some common hardwoods and softwoods (© BRE.)

3.2 Specific gravity

The traditional definition of specific gravity G can be expressed as

$$G = \frac{\rho_t}{\rho_w} \tag{3.5}$$

where ρ_t is the density of timber, and ρ_w is the density of water at 4 °C = 1.0000 g/cc. The value of G will therefore vary with moisture content and consequently the specific gravity of timber is usually based on the oven-dry mass, and volume at some specified moisture content. This is frequently taken as zero though, for convenience, green or other moisture conditions are sometimes used when the terms *basic specific gravity* and *nominal specific gravity* are applied respectively. Hence,

$$G_\mu = \frac{m_0}{V_\mu \rho_w} \tag{3.6}$$

where m_0 is the oven-dry mass of timber, V_μ is the volume of timber at moisture content μ, ρ_w is the density of water, and G_μ is the specific gravity at moisture content μ.

At low moisture contents, the specific gravity decreases slightly with increasing moisture content up to 30%, thereafter remaining constant. In research activities specific gravity is defined usually in terms of oven-dry mass and volume. However, for engineering applications specific gravity is frequently presented as the ratio of oven-dry mass to volume of timber at 12% moisture content. This can be derived from the oven-dry specific gravity, as

$$G_{12} = \frac{G_0}{1 + 0.01\mu G_0/G_{s12}} \tag{3.7}$$

where G_{12} is the specific gravity of timber at 12% moisture content, G_0 is the specific gravity of timber at zero moisture content, μ is the percentage moisture content, and G_{s12} is the specific gravity of bound water at 12% moisture content.

The relationship between density and specific gravity can be expressed as

$$\rho = G(1 + 0.01\mu)\rho_w \tag{3.8}$$

where ρ is the density at moisture content μ, G is the specific gravity at moisture content μ, and ρ_w is the density of water. Equation (3.8) is valid for all moisture contents. When $\mu = 0$ the equation reduces to

$$\rho = G_0 \tag{3.9}$$

i.e. the density and specific gravity are numerically equal.

3.3 Density of the dry cell wall

Although the density of timber may vary considerably among different timbers, the density of the actual cell wall material remains constant for all timbers with a value of approximately 1500 kg/m^3 (1.5 g/cc) when measured by volume-displacement methods.

The exact value for cell-wall density depends on the liquid used for measuring the volume. When a polar or swelling liquid such as water is used, the apparent specific volume of the dry cell wall is lower than when a non-polar (non-swelling) liquid such as toluene is used. Densities of 1.525 and 1.451 g/cc have been recorded for the same material using water and toluene respectively. This disparity in density of the dry cell wall has been explained in terms of, first, the greater penetration of a polar liquid into microvoid spaces in the cell wall which are inaccessible to non-polar liquids and, second, the compaction or apparent reduced volume of the sorbed water compared with the free-liquid water (Stamm, 1964). The voids are considered to occupy about 5% of the volume of the cell wall and consequently it has been shown that about 85% of the difference in specific volume of the dry cell wall (and hence density) by polar and non-polar liquid displacement is caused by the lower accessibility of the latter to the microvoids in the cell wall and only 15% of the difference is credited to the apparent compression of the sorbed water (Weatherwax and Tarkow, 1968).

Cell wall density can be measured by optical techniques as well as by volume displacement. Generally, density is calculated from cell wall measurements made on microtomed cross-sections, but the values obtained are usually lower than those by the volume-displacement method. The lack of agreement is usually explained in terms of either very fine damage to the cell wall during section cutting, or to inaccuracies in measuring the true cell wall thickness due to *shadow effects* resulting from the use of a point source of light (Petty, 1971). Papers published prior to 1966 have been extensively reviewed by Kellogg and Wangaard (1969).

3.4 Porosity

In Section 1.2.2 the cellular nature of timber was described in terms of a parallel arrangement of hollow tubes. The *porosity* (p) of timber is defined as the fractional void volume and is expressed mathematically as

$$p = 1 - V_f \tag{3.10}$$

where V_f is the volume fraction of cell wall substance.

Provided both the density of the cell wall substance and the moisture content of the timber are known, the volume fraction of the cell wall substance can be determined as

$$V_f = G\left(\frac{1}{\rho_c} + \frac{0.01}{G_s}\mu\right) \qquad (3.11)$$

where G is the specific gravity at moisture content μ, G_s is the specific gravity of bound water at moisture content μ, μ is the moisture content less than 25%, and ρ_c is the density of cell wall material (=1.46 when measured by helium displacement). Knowing V_f, the porosity can be calculated from equation (3.10).

A good approximation of p can be obtained if ρ_c is taken as 1.5 and G_s as 1.0; then,

$$p = 1 - G(0.667 + 0.01\ \mu) \qquad (3.12)$$

As an example of the use of equation (3.12), let us calculate the porosity of both balsa and lignum vitae, the densities of which are quoted in Section 3.1. Let $\mu = 0$, then $G_0 = \rho$ numerically. Volumetric shrinkages from 12% to 0% moisture content for balsa and lignum vitae are approximately 1.8% and 10%, respectively. From equation (3.4),

$$\rho_0 = \rho_{12}\frac{(1 + 0.01S_v)}{(1 + 0.01\mu)} \qquad (3.13)$$

Therefore, for balsa

$$\rho_0 = 176\frac{(1 + 0.018)}{(1 + 0.12)}$$

$$= 160 \text{ kg/m}^3$$

and, for lignum vitae,

$$\rho_0 = 1230\frac{(1 + 0.10)}{(1 + 0.12)}$$

$$= 1208 \text{ kg/m}^3$$

Hence, G_0 balsa = 0.160 and G_0 lignum vitae = 1.208. Therefore, from equation (3.12), the porosities p at zero moisture content are

balsa: 1 – 0.160(0.667) = 0.89 or 89%

lignum vitae: 1 – 1.208(0.667) = 0.19 or 19%

References

Kellogg, R.M. and Wangaard, F.F. (1969). Variation in the cell-wall density of wood. *Wood and Fiber*, **1** (3), 180–204.

Petty, J.A. (1971) The determination of fractional void volume in conifer wood by microphotometry, *Holzforschung*, **25** (1), 24–29.

Stamm, A.J. (1964) *Wood and Cellulosic Science*, Ronald Press, New York.

Weatherwax, R.C. and Tarkow, H. (1968) Density of wood substance: importance of penetration and adsorption compression of the displacement fluid, *For. Prod. J.*, **18** (7) 44–46.

Chapter 4

Movement in timber

4.1 Introduction

Timber in an unstressed state may undergo dimensional changes following variations in its moisture content and/or temperature. The magnitude, and consequently the significance, of such changes in the dimensions of timber is much greater in the case of alterations in moisture content compared with temperature. Although thermal movement will be discussed, the greater emphasis in this chapter is placed on the influence of changing moisture content.

4.2 Dimensional change due to moisture

4.2.1 Equilibrium moisture content

Timber is hygroscopic, that is it will absorb moisture from the atmosphere if it is dry and correspondingly yield moisture to the atmosphere when wet, thereby attaining a moisture content which is in equilibrium with the water vapour pressure of the surrounding atmosphere. Thus, for any combination of vapour pressure and temperature of the atmosphere there is a corresponding moisture content of the timber such that there will be no inward or outward diffusion of water vapour. This moisture content is referred to as the *equilibrium moisture content* or e.m.c. Generally, it is more convenient to use relative humidity rather than vapour pressure. Relative humidity is defined as the ratio of the partial vapour pressure in the air to the saturated vapour pressure, expressed as a percentage.

The fundamental relationships between moisture content of timber and atmospheric conditions have been determined experimentally and the average equilibrium moisture content values are shown graphically in Figure 4.1. A timber in an atmosphere of 20°C and 22% relative humidity will have a moisture content of 6% (see below), whereas the same timber if moved to an atmosphere of 40°C and 64% relative humidity will double its moisture content. It should be emphasised that the curves in Figure 4.1 are average values for moisture in relation to relative humidity and temperature, and that slight variations in the equilibrium moisture content will occur due to differences between timbers and to the previous history of the timber with respect to moisture.

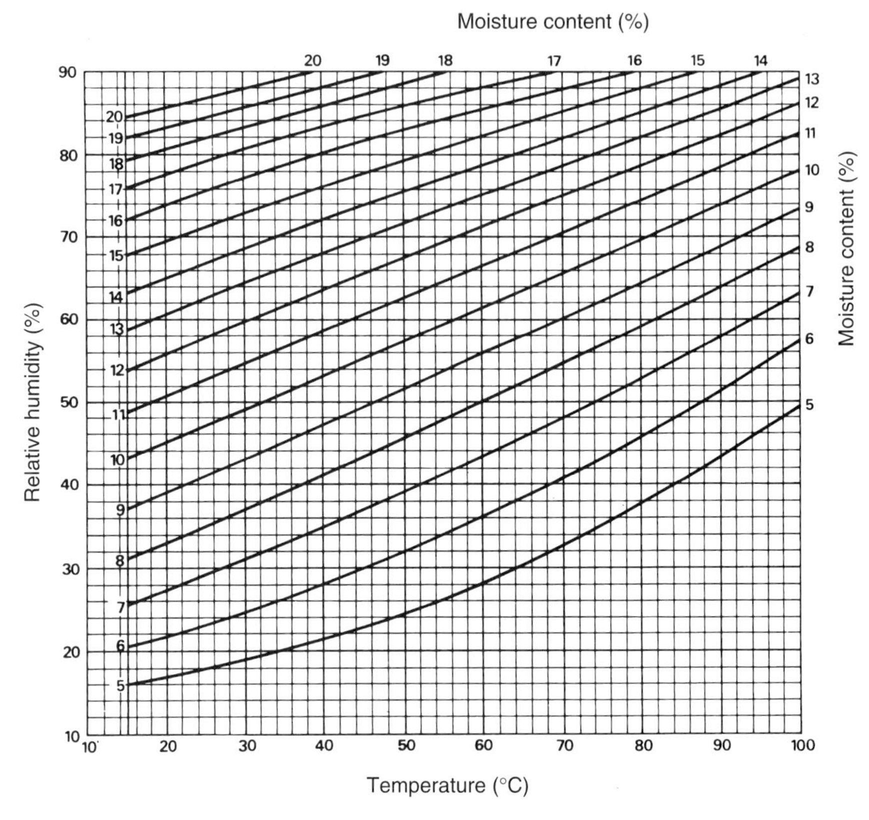

Figure 4.1 Chart showing the relationship between the moisture content of wood and the temperature and relative humidity of the surrounding air. The approximate curves are based on values obtained during drying from green condition. (© BRE.)

4.2.2 Determination of moisture content

It is customary to express the moisture content of timber in terms of its oven-dry mass using the equation

$$\mu = \frac{m_{\text{init}} - m_{\text{od}}}{m_{\text{od}}} \times 100 \tag{4.1}$$

where m_{init} is the initial mass of timber sample (grams), m_{od} is the mass of timber sample after oven-drying at 105 °C (grams) and μ is the moisture content of timber sample (%).

The expression for the moisture content of timber on a dry-mass basis is in contrast to the procedure adopted for other materials where moisture content is expressed in terms of the wet mass of the material.

Determination of moisture content in timber is usually carried out using the basic gravimetric technique above, though it should be noted that at least a dozen different methods have been recorded in the literature. Suffice it here to mention only two of these alternatives. First, where the timber contains volatile extractives which would normally be lost during oven drying, thereby resulting in erroneous moisture content values, it is customary to use a distillation process, heating the timber in the presence of a water-immiscible liquid such as toluene, and collecting the condensed water vapour in a calibrated trap. Second, where ease and speed of operation are preferred to extreme accuracy, moisture contents are assessed using electric moisture meters. The type most commonly used is known as the *resistance meter*, though this battery-powered hand-held instrument actually measures the conductance or flow (the reciprocal of resistance) of an electric current between two probes. Below the fibre saturation point (about 27% moisture content, Section 4.2.6) an approximately linear relationship exists between the logarithm of conductance and the logarithm of moisture content. However, this relationship, which forms the basis for this type of meter, changes with species of timber, temperature and grain angle. Thus, a resistance-type meter is equipped with a number of alternative scales, each of which relates to a different group of timber species; it should be used at temperatures close to 20 °C with the pair of probes inserted parallel to the grain direction (the method is described more fully in EN 13183–2).

Although the measurement of moisture content is quick with such a meter, there are three drawbacks to its use. First, moisture content is measured only to the depth of penetration of the two probes, a measurement which may not be representative of the moisture content of the entire depth of the timber member – the use of longer probes can be beneficial though these are difficult to insert and withdraw. Second, the moisture content obtained is not so accurate as that obtained from the gravimetric method described previously. Third, the working range of the instrument is only from 7% to 27% moisture content. The upper limit is determined by the onset of non-linearity in the basic relationship between conductance and moisture content, whereas the lower limit results from the difficulty of measuring the very small conductance at these low levels of moisture content.

Other types of meter are based either on determining the dielectric constant of timber (capacitance type meter), or the dielectric loss factor (power-loss type). These meters have the advantage over the resistance type in that they use a plate applied to the timber surface rather than the insertion of probes into the timber. However, dielectric meters are influenced by timber density and corrections have to be made to the readings obtained. Like the resistance meter, they only give mean readings of moisture content and cannot determine the moisture gradient. Additionally, the working range of the dielectric meter is reduced to 15–27% moisture content and, therefore, it is not surprising that few dieletric-based meters are commercially available.

Further information on these three types of meters and their correct use is given in the comprehensive review by James (1988).

4.2.3 *Moisture content of green timber*

In the living tree, water is to be found not only in the cell cavity, but also within the cell wall. Consequently the moisture content of green wood (newly felled) is high, usually varying from about 60% to nearly 200% depending on the location of the timber in the tree and the season of the year. However, seasonal variation is slight compared to the differences that occur within a tree between the sapwood and heartwood regions. The degree of variation is illustrated for a number of softwoods and hardwoods in Table 4.1. Within the softwoods the sapwood may contain twice the percentage of moisture to be found in the corresponding heartwood, whereas in the hardwoods this difference is appreciably smaller or even absent. However, pockets of 'wet' wood can be found in the heartwood as described in Section 1.2.1.

Green timber will yield moisture to the atmosphere with consequent changes in its dimensions. At moisture contents above 20% many timbers, especially their sapwood, are susceptible to attack by fungi; the strength and stiffness of green wood is considerably lower than for the same timber when dry. For all these reasons it is necessary to dry or *season* timber following felling of the tree and prior to its use in service.

4.2.4 *Removal of moisture from timber*

Drying or seasoning of timber can be carried out in the open, preferably with a top cover. However, it will be appreciated from the previous discussion on equilibrium moisture contents that the minimum moisture content that can be achieved is determined by the lowest relative humidity of the summer period.

Table 4.1 Average green moisture contents of the sapwood and heartwood

		Moisture content	
Botanical name	*Commercial name*	*Heartwood (%)*	*Sapwood (%)*
Hardwoods			
Betula lutea	Yellow birch	64	68
Fagus grandifolia	American beech	58	79
Ulmus americana	American elm	92	84
Softwoods			
Pseudotsuga menziesii	Douglas fir	40	116
Tsuga heterophylla	Western hemlock	93	167
Picea sitchensis	Sitka spruce	50	131

In this country it is seldom possible to achieve moisture contents less than 16% by air seasoning. The planks of timber are separated in rows by stickers (usually 25–30 mm across) which permit air currents to pass through the pile; nevertheless it may take from 2 to 10 years to air-season timber depending on the species of timber and the thickness of the timber members.

The process of seasoning may be accelerated artificially by placing the stacked timber in a drying kiln, basically a large chamber in which the temperature and humidity can be controlled and altered throughout the drying process; the control may be carried out manually or programmed automatically. Humidification is sometimes required in order to keep the humidity of the surrounding air at a desired level when insufficient moisture is coming out of the timber; it is frequently required towards the end of the drying run and is achieved either by the admission of small quantities of live steam or by the use of water atomisers or low-pressure steam evaporators. Various designs of kilns are used and these are reviewed in detail by Pratt (1974).

Drying of softwood timber in a kiln can be accomplished in from 4 to 7 days, the optimum rate of drying varying widely from one timber to the next; hardwood timber usually takes about three times longer than softwood of the same dimension. Following many years of experimentation, kiln schedules have been published for different groups of timbers. These schedules provide wet- and dry-bulb temperatures (maximum of 70 °C) for different stages in the drying process and their use should result in the minimum amount of degrade in terms of twist, bow, spring, collapse and checks (Pratt, 1974). Most timber is now seasoned by kilning; little air drying is carried out. Dry stress-graded timber in the UK must be kiln-dried to a mean value of 20% moisture content with no single piece greater than 24%. However, UK and some Swedish mills are now targeting 12% ('superdried'), as this level is much closer to the moisture content in service.

Recently, *solar kilns* have become commercially available and are particularly suitable for use in the developing countries to season many of the difficult slow-drying tropical timbers. These small kilns are very much cheaper to construct than conventional kilns and are also much cheaper to run. They are capable of drying green timber to about 7% moisture content in the dry season and about 11% in the rainy season.

Another recent innovation in the seasoning of timber is the use of *high-temperature drying*. This process is used commercially in Australia and New Zealand for the rapid drying of *Pinus radiata* timber. The aim is to dry a stack of timber within a 24 h period by initially boiling off the water for the first 12–15 h at a temperature of 120 °C, followed by a period (9–12 h) of cooling and reconditioning to the desired moisture content. Heavy top loading of the stack is required in order to restrain the amount of bow, twist and collapse of the timber. Early work on spruce in the UK has been unsuccessful in that much case-hardening occurred; this is probably in part a reflection of the much lower permeability of the timber of this genus.

4.2.5 *Influence of structure*

As mentioned in Section 4.2.3, water in green or freshly felled timber is present both in the cell cavity and within the cell wall. During the seasoning process, irrespective of whether this is by air or within a kiln, water is first removed from within the cell cavity: this holds true down to moisture contents of about 27–30%. As the water in the cell cavities is *free*, not being chemically bonded to any part of the timber, it can readily be appreciated that its removal will have no effect on the strength or dimensions of the timber. The lack of variation of the former parameter over the moisture content range of 110–27% is illustrated in Figure 4.2.

However, at moisture contents below 27–30% water is no longer present in the cell cavity but is restricted to the cell wall where it is chemically bonded (hydrogen bonding) to the matrix constituents, to the hydroxyl groups of the cellulose molecules in the noncrystalline regions and to the surface of the

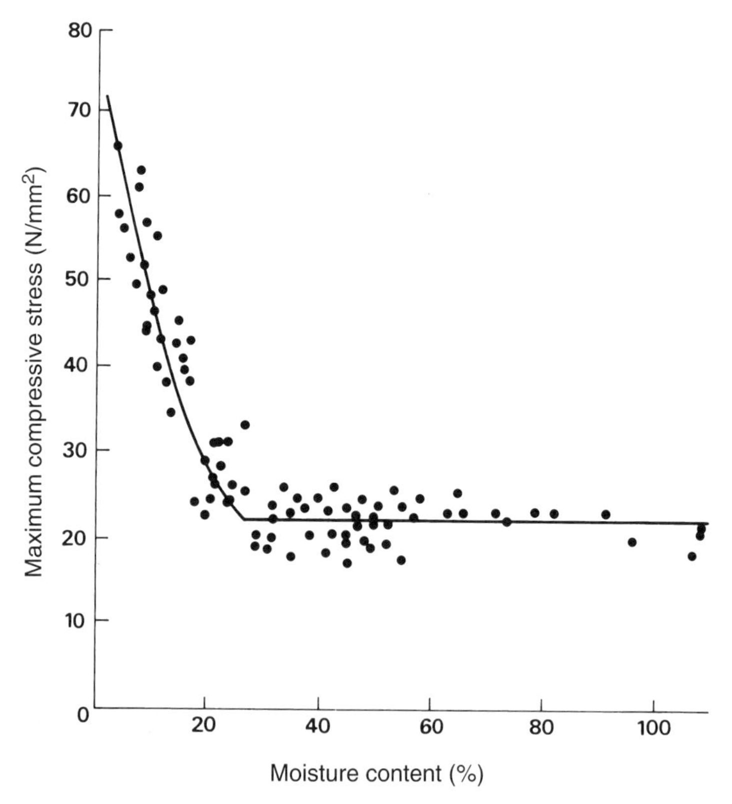

Figure 4.2 Relationship between longitudinal compressive strength and moisture content. (© BRE.)

crystallites; as such, this water is referred to as *bound water*. The uptake of water by the lignin component is considerably lower than that by either the hemicellulose or the amorphous cellulose. As discussed in Section 4.2.7, water may be present as a monomolecular layer though frequently up to six layers can be present. Water cannot penetrate the crystalline cellulose because the hygroscopic hydroxyl groups are mutually satisfied by the formation of both intra- and intermolecular bonds within the crystalline region as described in Chapter 1. This view is confirmed from X-ray analyses which indicate no change of state of the crystalline core as the timber gains or loses moisture.

However, the percentage of noncrystalline material in the cell wall varies between 8% and 33%, and the influence of this fraction of cell wall material as it changes moisture content on the behaviour of the total cell wall is very significant. The removal of water from these areas within the cell wall results first in increased strength and, second, in marked shrinkage. Both changes can be accounted for in terms of drying out of the water-reactive matrix, thereby causing the microfibrils to come into closer proximity, with a commensurate increase in interfibrillar bonding and decrease in overall dimensions. Such changes are reversible, or almost completely reversible.

4.2.6 Fibre saturation point

The increase in strength on drying is clearly indicated in Figure 4.2, from which it will be noted that there is a threefold increase in strength as the moisture content of the timber is reduced from about 27% to zero. The moisture content corresponding to the inflexion in the graph is termed the *fibre saturation point* (f.s.p.), where in theory there is no free water in the cell cavities while the walls are holding the maximum amount of bound water. In practice this rarely exists – a little free water may still exist while some bound water is removed from the cell wall. Consequently, the fibre saturation point, although a convenient concept, should really be regarded as a 'range' in moisture contents over which the transition occurs.

The fibre saturation point therefore corresponds in theory to the moisture content of the timber when placed in a relative humidity of 100%. In practice, however, this is not so because such an equilibrium would result in total saturation of the timber (Stamm, 1964). Values of e.m.c. above 98% are unreliable. It is generally found that the moisture content of hardwoods at this level are from 1% to 2% higher than for softwoods. At least nine different methods of determining the fibre saturation point are recorded in the literature. These include the extrapolation of moisture content absorption isotherms to unit relative vapour pressure (now considered to be inaccurate), the non-solvent water method, the polymer-exclusion method, the pressure plate system, and the determination of the inflection point in plotting the logarithm of electrical conductivity against moisture content. It must be appreciated that the value of the fibre saturation point is dependent on the method used.

4.2.7 Sorption and diffusion

As noted in Section 4.2.1, timber assumes with the passage of time a moisture content in equilibrium with the relative vapour pressure of the atmosphere. This process of water sorption is typical of solids with a complex capillary structure and this phenomenon has also been observed in concrete. The similarity in behaviour between timber and concrete with regard to moisture relationships is further illustrated by the presence of S-shaped isotherms when moisture content is plotted against relative vapour pressure. Both materials have isotherms which differ according to whether the moisture content is reducing (desorption) or increasing (adsorption) thereby producing a *hysteresis loop* (Figure 4.3).

The *hysteresis coefficient*, *A*/*D*, is defined as the ratio of the equilibrium moisture content for adsorption to that for desorption for any given relative vapour pressure. For timber this varies from 0.8 to 0.9 depending on species of timber and on ambient temperature. The effect of increasing temperature on reducing the equilibrium moisture content of timber is illustrated in Figure 4.4.

The hysteresis effect in sorption was at first explained in terms of differences in contact angle of the advancing and receding water front within the cell cavities, which are considered to be capillaries. Later, the hysteresis effect was related to the behaviour of the hydroxyl groups in both the cellulose and lignin in drying these groups are thought to satisfy each other and on rewetting; some continue to do so and are not available for water adsorption. According to a

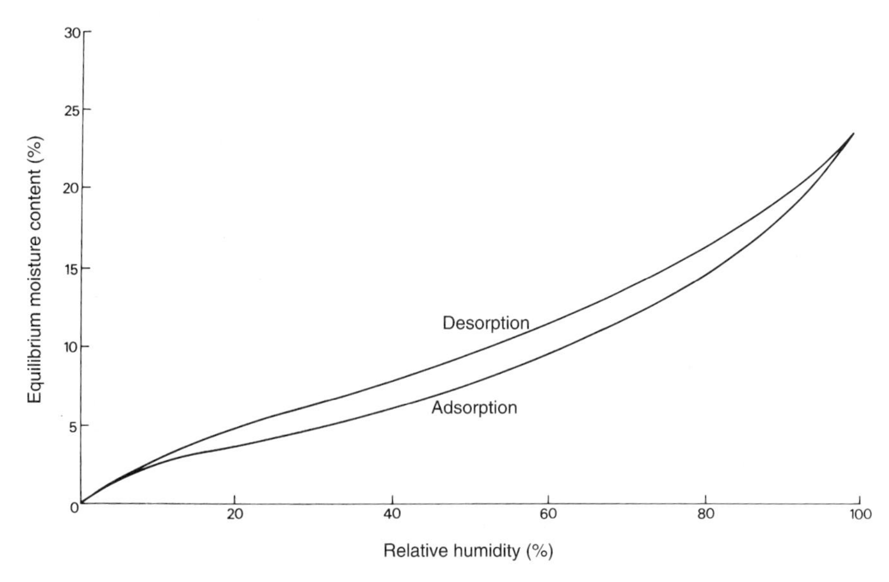

Figure 4.3 Hysteresis loop resulting from the average adsorption and desorption isotherms for six species of timber at 40 °C (©BRE.)

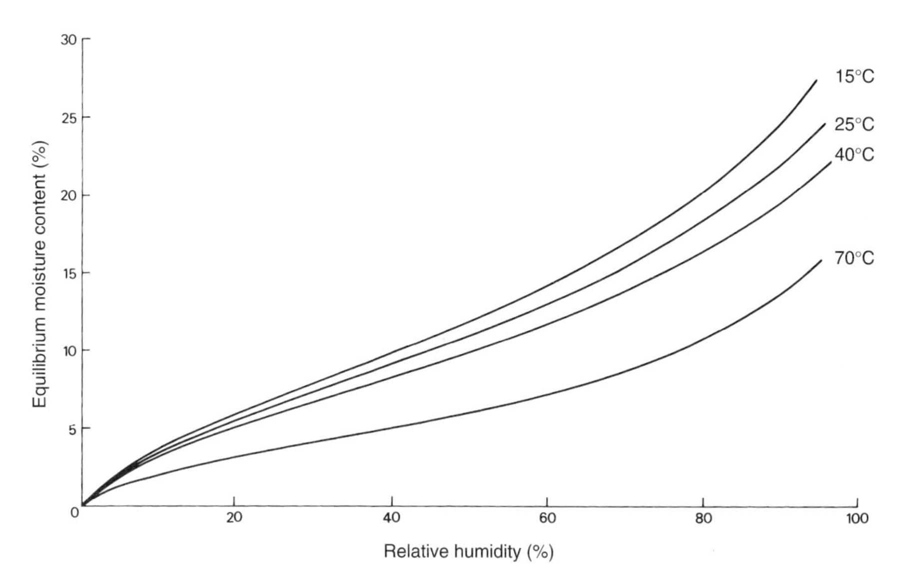

Figure 4.4 Average desorption isotherms for six species at four temperatures. (© BRE.)

third hypothesis, plasticity is considered to be the principal cause of hysteresis and is related to irreversible inelastic exchanges of hydroxyl groups between neighbouring cellulose molecules (Barkas, 1949).

Much more attention has been paid over the years to the development of sorption theories in general, than to an explanation of the hysteresis behaviour in particular. The sorption of water by timber is a surface phenomenon and explainable in terms of the hydroxyl groups that are mostly located in the amorphous areas of the cellulose and the hemicelluloses, and which display a great affinity for water.

In timber with a moisture content less than 6%, adsorption of water is by chemisorption (primary water) to produce a monolayer of water molecules on the wood surface. At moisture contents between 6% and 15%, a multilayer of water molecules is formed by the adsorption of secondary water such that five or six water molecules can be attracted to each adsorption site. Above 15% moisture content, adsorption of water occurs by the formation of water clusters which, above 20% moisture content, may average more than 10 molecules (Avramidis, 1997). It is possible to relate these occurrences to the three parts of the sigmoid-shaped isotherm.

The adsorption of water on wood surfaces is always accompanied by the evolution of heat and recently a number of attempts have been made to obtain an effective and representative thermodynamic analysis of heat of sorption data. Because of the existence of sorption hysteresis, the wood–water system is not truly reversible and strictly speaking classical thermodynamics is not applicable. Nevertheless, it has been applied in order to calculate approximate magnitudes of a number of thermodynamic parameters such as changes in enthalpy and free energy during moisture sorption; many of the sorption theories present in the literature embody sorption energies and water adsorption in timber does appear to be enthalpy driven (Avramidis, 1997).

Several moisture sorption theories have been proposed over the years. Perhaps the best known is the Brunauer–Emmet–Teller (BET) theory (Brunauer *et al.* 1938) which assumes that from one to many layers of water molecules are present on the sorption sites. This theory was later modified by Dent (1977) to provide a better description of the sorption isotherm of wood. Other theories which have provided a good fit to the sigmoid sorption isotherm for timber are those by Hailwood and Horrobin (1946) and by Anderson and McCarthy (1963). More information on the important areas of thermodynamics and sorption theories for water in timber is given in the comprehensive accounts by Simpson (1980), Skaar (1988), Siau (1995) and Avramidis (1997).

4.2.8 Dimensional changes

In timber it is customary to distinguish between those changes that occur when green timber is dried to very low moisture contents (e.g. 12%), and those that arise in timber of low moisture content due to seasonal or daily changes in the relative humidity of the surrounding atmosphere. The former changes are called *shrinkage*, whereas the latter are known as *movement*.

4.2.8.1 Shrinkage

ANISOTROPY IN SHRINKAGE

The reduction in dimensions of the timber, technically known as *shrinkage,* can be considerable but, owing to the complex structure of the material, the degree of shrinkage is different on the three principal axes; in other words, timber is anisotropic in its water relationships. The variation in degree of shrinkage that occurs between different timbers and, more important, the variation among the three different axes within any one timber is illustrated in Table 4.2. It should be noted that the values quoted in the table represent shrinkage on drying from the green state (i.e. above 27%) to 12% moisture content, a level which is of considerable practical significance. At 12% moisture content, timber is in equilibrium with an atmosphere having a relative humidity of 60% and a temperature of 20 °C; these conditions would be found in buildings having regular, but intermittent, heating.

From Table 4.2 it will be observed that shrinkage varies from 0.1% to 10%, i.e. over a 100-fold range. Longitudinal shrinkage, is always an order of magnitude less than transverse, whereas in the transverse plane radial shrinkage is usually some 60–70% of the corresponding tangential figure.

Table 4.2 Shrinkage (%) on drying from green to 12% moisture content

		Transverse		
Botanical name	*Commercial name*	*Tangential*	*Radial*	*Longitudinal*
Chlorophora excelsa	Iroko	2.0	1.5	<0.1
Tectona grandis	Teak	2.5	1.5	<0.1
Pinus strobus	Yellow pine	3.5	1.5	<0.1
Picea abies	Whitewood	4.0	2.0	<0.1
Pinus sylvestris	Redwood	4.5	3.0	<0.1
Tsuga heterophylla	Western hemlock	5.0	3.0	<0.1
Quercus robur	European oak	7.5	4.0	<0.1
Fagus sylvatica	European beech	9.5	4.5	<0.1

The anisotropy between longitudinal and transverse shrinkage, amounting to approximately 40:1, is due in part to the vertical arrangement of cells in timber and in part to the particular orientation of the microfibrils in the middle layer of the secondary cell wall (S_2).

As the microfibrils of the S_2 layer of the cell wall are inclined at an angle of about 15° to the vertical, the removal of water from the matrix and the consequent movement closer together of the microfibrils will result in a horizontal component of the movement considerably greater than the corresponding longitudinal component. This is apparent in practice (see Table 4.2) though the relationship between shrinkage and microfibrillar angle is non-linear (Figure 4.5). The variation in shrinkage with angle for tangential shrinkage is, as might be expected, almost the mirror image of that for longitudinal shrinkage. The cross-over point, when the degree of anisotropy is zero, occurs experimentally at a microfibrillar angle of about 48°. Although the microfibrillar angle is of supreme importance, it is apparent that the relationship is complex and that other factors, as yet unknown, are playing additional though minor roles.

Various theories have been developed over the years to account for shrinkage in terms of microfibrillar angle. The early theories were based on models which generally consider the cell wall to consist of an amorphous hygroscopic matrix in which are embedded parallel crystalline microfibrils which restrain swelling or shrinking of the matrix. One of the first models considered part of the wall as a flat sheet consisting only of an S_2 layer in which the microfibrillar angle had a constant value (Barber and Meylan, 1964). This model treated the cells as square in cross-section and there was no tendency for the cells to twist as they began to swell. An improved model (Barber, 1968) was to treat the cells as circular in cross-section and embraced a thin constraining sheath outside the main cylinder which acted to reduce transverse swelling. Experimental confirmation of this model by Meylan (1968) is illustrated in Figure 4.5 in which the calculated lines from the model show good agreement with the experimentally derived points. Later models have treated the cell wall as two layers

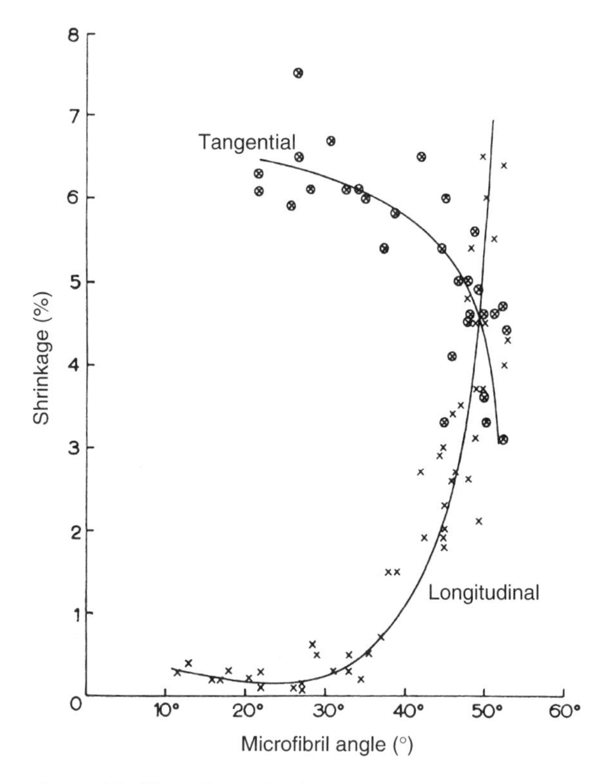

Figure 4.5 The relationship between longitudinal and tangential shrinkage and microfibril angle of the S_2 layer in *Pinus jeffreyi*. (From Meylan (1968) *Forest Products Journal*, **18** (4), 75–78, reproduced by permission of the Forest Products Society.)

of equal thickness having microfibrillar angles of equal and opposite sense, and these two-ply models have been developed extensively over the years to take into account the layered structured of the cell wall, differences in structure between radial and tangential walls, and variations in wall thickness. The principal researcher using this later type of model was Cave. His models are based on an array of parallel cellulose microfibrils embedded in a hemicellulose matrix, with different arrays for each wall layer. These arrays of basic wall elements were bonded together by lignin microlayers. Earlier versions of the model included consideration of the variation in the stiffness of the matrix with changing moisture content (Cave 1972, 1975). The model was later modified to take account of the amount of high-energy water absorbed rather than the total amount of water (Cave, 1978a,b). Comparison with previously obtained experimental data was excellent at low moisture contents, but poorer at moisture contents between 15% and 25%. All these theories are extensively presented and discussed by Skaar (1988).

The influence of microfibrillar angle on the degree of longitudinal and transverse shrinkage described for normal wood is supported by evidence derived from experimental work on *compression wood*, one of the forms of reaction wood described in Chapter 1. Compression wood is characterised by possessing a middle layer to the cell wall, the microfibrillar angle of which can be as high as 45°, although 20–30° is more usual. The longitudinal shrinkage is much higher and the transverse shrinkage correspondingly lower than in normal wood and it has been demonstrated that the values for compression wood can be accommodated on the shrinkage–angle curve for normal wood.

Differences in the degree of transverse shrinkage between tangential and radial planes (Table 4.2), are usually explained in the following terms. First, the restricting effect of the rays on the radial plane; second, the increased thickness of the middle lamella on the tangential plane compared with the radial; third, the difference in degree of lignification between the radial and tangential cell walls; fourth, the small difference in microfibrillar angle between the two walls; and fifth, the alternation of earlywood and latewood in the radial plane, which, because of the greater shrinkage of latewood, induces the weaker earlywood to shrink more tangentially than it would if isolated. This last-mentioned explanation is known as the *Mörath* or *earlywood–latewood interaction theory*. Considerable controversy reigns as to whether all five factors are actually involved and their relative significance. Comprehensive reviews of the evidence supporting these five possible explanations of differential shrinkage in the radial and tangential planes is to be found in Pentoney (1953), Boyd (1974) and Skaar (1988).

Kifetew (1997) has demonstrated that the gross transverse shrinkage anisotropy of Scots pine timber, with a value approximately equal to 2, can be explained primarily in terms of the earlywood–latewood interaction theory by using a set of mathematical equations proposed by him; the gross radial and transverse shrinkage values were determined from the isolated early and latewood shrinkage values taken from the literature.

The volumetric shrinkage, s_v is slightly less than the sums of the three directional components and is given by

$$s_v = 100[1 - (1 - 0.01s_l)(1 - 0.01s_r)(1 - 0.01s_t)] \qquad (4.2)$$

where the shrinkages are in percentages. This equation simplifies to

$$s_v = s_l + s_r + s_t - 0.01 s_r s_t \qquad (4.3)$$

and subsequently to

$$s_v = s_r + s_t \qquad (4.4)$$

as greater approximations become acceptable.

PRACTICAL SIGNIFICANCE.

In order to avoid the shrinkage of timber after fabrication, it is essential that it is dried down to a moisture content which is in equilibrium with the relative humidity of the atmosphere in which the article is to be located. A certain latitude can be tolerated in the case of timber frames and roof trusses, but in the production of furniture, window frames, flooring and sports goods it is essential that the timber is seasoned to the expected equilibrium conditions, namely 12% for regular intermittent heating and 10% in buildings with central heating, otherwise shrinkage in service will occur with loosening of joints, crazing of paint films, and buckling and delamination of laminates. An indication of the moisture content of timber used in different environments is presented in Figure 4.6.

4.2.8.2 Movement

So far only those dimensional changes associated with the initial reduction in moisture content have been considered. However, dimensional changes, albeit smaller in extent, can also occur in seasoned or dried wood due to changes in the relative humidity of the atmosphere. Such changes certainly occur on a seasonal basis and frequently also on a daily basis. As these changes in humidity are usually fairly small, inducing only slight changes in the moisture content of the timber, and as a considerable delay occurs in the diffusion of water vapour into or out of the centre of a piece of timber, it follows that these dimensional changes in seasoned timber are small, considerably smaller than those for shrinkage.

To quantify such movements for different timbers, dimensional changes are recorded over an arbitrary range of relative humidities. In the UK, the standard procedure is to condition the timber in a chamber at 90% relative humidity and 25 °C, then to measure its dimensions and to transfer it to a chamber at 60% relative humidity, allowing it to come to equilibrium before remeasuring it; the corresponding average change in moisture content is from 21% to 12%. Movement values in the tangential and radial planes for those timbers listed in Table 4.2 are presented in Table 4.3. The timbers are recorded in the same order, thus illustrating that although a broad relationship holds between values of shrinkage and movement, individual timbers can behave differently over the reduced range of moisture contents associated with movement. As movement in the longitudinal plane is so very small, it is generally ignored. Anisotropy within the transverse plane can be accounted for by the same set of variables that influence shrinkage.

Where timber is subjected to wide fluctuations in relative humidity, care must be exercised to select a species that has low movement values.

Moisture in timber has a very pronounced effect not only on its strength (Figure 4.2), but also on its stiffness, toughness and fracture morphology. These aspects are discussed in Chapters 6 and 7.

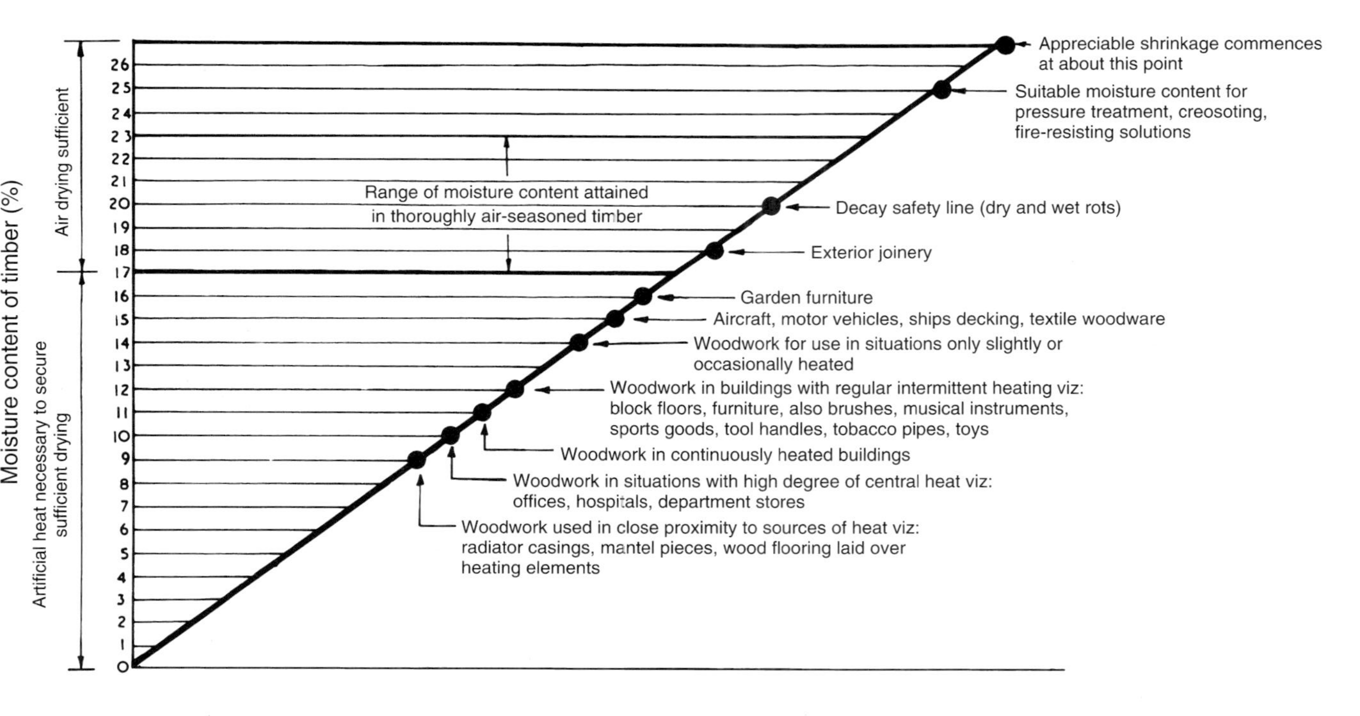

Figure 4.6 Equilibrium moisture content of timber in various environments. The figures for different species vary, and the chart shows only average values. (© BRE.)

Table 4.3 Movement (%) on transferring timber from 90% relative humidity to 60% at 25 °C

Botanical name	Commercial name	Transverse	
		Tangential	Radial
Chlorophora excelsa	Iroko	1.0	0.5
Tectona grandis	Teak	1.2	0.7
Pinus strobus	Yellow pine	1.8	0.9
Picea abies	Whitewood	1.5	0.7
Pinus sylvestris	Redwood	2.2	1.0
Tsuga heterophylla	Western hemlock	1.9	0.9
Quercus robur	European oak	2.5	1.5
Fagus sylvatica	European beech	3.2	1.7

4.2.9 Dimensional stabilisation

Aware of the technological significance of the instability of wood under changing moisture content, many attempts have been made over the years to find a solution to the problem.

The production of plywood from timber by slicing into thin veneers which are then laid up such that alternate veneers are at right angles to each other does much to reduce the considerable horizontal movement of timber, though the movement of plywood in any direction is still higher than the original longitudinal movement of the timber from which it was made.

The application of various oil-based or alkyl-based finishes result in a marked reduction in the movement of the timber substrate, certainly in the short term. However, in prolonged exposure to high humidity, movement of the substrate will occur, albeit at a slower rate than in the absence of a coating, as no commercially available synthetic finish is completely impermeable to water vapour.

Good dimensional stabilisation can be obtained, however, by heating the timber for short periods of time to very high temperatures (250–350 °C); a reduction of 40% has been recorded after heating timber to 350 °C for short periods of time (Rowell and Youngs, 1981). It is possible to achieve a reduction in swelling of from 50% to 80% at lower temperatures (180–200 °C), again in short periods of time, by heating the sample in an inert gas at 8–10 bar (Giebeler, 1983). The use of even lower temperatures (120–160 °C) in the presence of air necessitates exposure for several months in order to achieve a similar reduction in swelling. Unfortunately, all these treatments result in thermal degradation of the timber with considerable loss in strength and especially toughness. It appears that it is the degradation of the hemicelluloses that reduces the propensity of the timber to swell (Stamm, 1977).

Since the late 1970s, research has shown that it is possible to reduce the hygroscopicity of timber through chemical substitution of the active hydroxyl groups by less polar groups. Greatest success has been achieved by acetylation of the timber; acetic anhydride is used as a source of acetyl groups with pyridine as a catalyst. A marked reduction in dimensional instability is obtained with only a marginal loss in strength (Rowell, 1984), but with a significant loss in tensile modulus (Ramsden *et al.*, 1997).

Two other chemical treatments which result in improved stability are cross-linking and wall bulking. In the former treatment, the timber is reacted with formaldehyde under acidic conditions with the formation of methylene bridges between adjacent hydroxyl groups; the process does require the timber to have a moisture content less than 10%. There is little gain in mass and there is no bulking of the cell wall (Stamm, 1977).

The other group of chemical treatments all result in bulking of the cell wall, thereby holding the timber in a swollen state and minimising further dimensional movement. Both polyethylene glycol (PEG) and phenol formaldehyde have been shown to give good results. PEG is usually applied to green wood where the PEG is exchanged for the cell wall water. It is also widely used in the conservation of previously waterlogged items. The PEG is not bonded to the wood and is leachable.

The impregnation of dry timber with phenol formaldehyde also results in the absence of chemical bonding to the cell wall, but has the advantage that it is non-leachable. Various commercial products are available, some of which have been subjected to compression before curing of the phenol formaldehyde.

Through chemical reactions it is possible to add an organic chemical to the hydroxyl groups which not only bulks up the cell wall, but is non-leachable. Various epoxides, anhydrides and isocyanates have all been used succcssfully in the laboratory to give very good stability; unfortunately, their cost tends to preclude most of them from commercial use.

Chemical dimensional stabilisers are discussed further in Section 9.3.1.3.

4.3 Thermal movement

Timber, like other materials, undergoes dimensional changes commensurate with increasing temperature. This is attributed to the increasing distances between the molecules as they increase the magnitude of their oscillations with increasing temperature. Such movement is usually quantified for practical purposes as the coefficient of linear thermal expansion and values for certain timbers are listed in Table 4.4. Although differences occur between species, these appear to be smaller than those occurring for shrinkage and movement. The coefficient for transverse expansion is an order of magnitude greater than that in the longitudinal direction. This degree of anisotropy (10:1) can be related to the ratio of length to breadth dimensions of the crystalline regions within the cell wall. Transverse thermal expansion appears to be correlated with specific gravity, but somewhat surprisingly this relationship is not sustained in the case of longitudinal thermal expansion where the values for different timbers are roughly constant (Weatherwax and Stamm, 1946).

Table 4.4 Coefficient of linear thermal expansion of various woods and other materials per degree Centigrade

Material	Coefficient of thermal expansion × 10^{-6}		
	Longitudinal		*Transverse*
Picea abies (Whitewood)	5.41		34.1
Pinus strobus (Yellow pine)	4.00		72.7
Quercus robur (European oak)	4.92		54.4
GRP, 60/40 unidirectional	10.0		10.0
CFRP, 60/40 unidirectional	10.0		−1.00
Mild steel		12.6	
Duralumin (an aluminium alloy)		22.5	
Nylon 66		125.0	
Polypropylene		110.0	

The expansion of timber with increasing temperature appears to be linear over a wide temperature range. The slight differences in expansion that occur between the radial and tangential planes are usually ignored and the coefficients are averaged to give a transverse value, as recorded in Table 4.4. For comparative purposes, the coefficients of linear thermal expansion for glass-reinforced plastic (GRP) and carbonfibre-reinforced plastic (CFRP), two metals and two plastics are also listed. Even the transverse expansion of timber is considerably less than that for the plastics.

The dimensional changes of timber caused by differences in temperature are small when compared to changes in dimensions resulting from the uptake or loss of moisture. Thus for timber with a moisture content greater than about 3%, the shrinkage due to moisture loss on heating will be greater than the thermal expansion, with the effect that the net dimensional change on heating will be negative. For most practical purposes thermal expansion or contraction can be safely ignored over the range of temperatures in which timber is generally employed.

References

Standards and specifications

EN 13183–2 *Round and sawn timber – Method of assessment of moisture content – Part 2: Method for estimating moisture content of a piece of sawn timber (Electrical method).*

Literature

Anderson N.T. and McCarthy J.L. (1963) Two parameter isotherm equation for fiber–water systems. *Ind. Eng. Chem. Process Design Dev.*, **2**, 103–105.

Avramidis, S. (1997) The basics of sorption, in *Proceedings of International conference on wood-water relations*, Copenhagen, Ed. P. Hoffmeyer, and published by the management committee of EC COST Action E8, 1–16.

Barber, N.F. (1968) A theoretical model of shrinking wood. *Holzforschung*, **22**, 97–103.

Barber, N.F. and Meylan, B.A. (1964) The anisotropic shrinkage of wood. A theoretical model. *Holzforschung*, **18**, 146–156.

Barkas, W.W. (1949) *The Swelling of Wood under Stress*, HMSO, London.

Boyd, J.D. (1974). Anisotropic shrinkage of wood. Identification of the dominant determinants. *Mokuzai Gakkaishi*, **20**, 473–482.

Brunauer, S.P., Emmett, P.H. and Teller, E. (1938) Adsorption of gases in multimolecular layers, *J. Am. Chem. Soc.*, **60**, 309–319.

Cave, I.D. (1972) A theory of shrinkage of wood. *Wood Sci. Technol.*, **6**, 284–292.

Cave, I.D. (1975) Wood substance as a water-reactive fibre reinforced composite. *J Microscopy*, **104**, 47–52.

Cave, I.D. (1978a) Modelling moisture-related mechanical properties of wood. I. Properties of the wood constituents. *Wood Sci. and Technol.*, **12**, 75–86.

Cave, I.D. (1978b) Modelling moisture-related mechanical properties of wood. II. Computation of properties of a model of wood and comparison with experimental data. *Wood Sci. Technol.*, **12**, 127–139.

Dent, R.W. (1977) A multilayer theory for gas sorption. I. Sorption of a single gas. *Text. Res. J.*, **47**, 145–152.

Giebeler, E.(1983) Dimensionsstabilisierung von holzduruch eine feuchte/warme/druckbehandlung. *Holz als Roh-und Werkstoff*, **41**, 87–94.

Hailwood, A.J. and Horrobin, S. (1946) Absorption of water by polymers; analysis in terms of a simple model. *Trans. Faraday Soc.*, **42B**, 84–102.

James, W.L. (1988) *Electric moisture meters for wood*, General Technical Report **6**, US Forest Products Laboratory, Madison.

Kifetew, G (1997) Application of the early–latewood interaction theory of the shrinkage anisotropy of Scots pine, *Proceedings of International conference on wood-water relations*, Copenhagen, Ed. P. Hoffmeyer, and published by the management committee of EC COST Action E8, 165–171.

Meylan, B.A. (1968) Cause of high longitudinal shrinkage in wood, *For. Prod. J.*, **18** (4), 75–78.

Pentoney, R.E. (1953) Mechanisms affecting tangential vs. radial shrinkage, *J. For. Prod. Res. Soc.*, **3**, 27–32.

Pratt, G.H. (1974) *Timber drying manual*. HMSO, London.

Ramsden, M.J, Blake, F.S.R. and Fey, N.J. (1997) The effect of acetylation on the mechanical properties, hydrophobicity, and dimensional stability of Pinus sylvestris. *Wood Sci. Technol.*, **31**, 97–104.

Rowell, R. M. (1984) Chemical modification of wood. *For. Prod. Abstracts*, **6**, 362–382.

Rowell, R.M. and Youngs R.L. (1981) *Dimensional stabilisation of wood in use*, Research Note FPL–0243, US Forest Products Laboratory, Madison.

Siau, J. (1995) *Wood: influence of moisture on physical properties*, VPI and SU Press, Blacksburg, VA.

Simpson, W. (1980) Sorption theories applied to wood. *Wood & Fiber*, **12** (3), 183–195.

Skaar, C. (1988) *Wood–water relations*, Springer-Verlag, Berlin.

Stamm, A.J. (1964) *Wood and cellulose science*, Ronald, New York.

Stamm, A.J. (1977) Dimensional changes of wood and their control, in I.S. Goldstein. (Ed.) *Wood Technology: Chemical Aspects*, A.C.S. Symposium Series 4, American Chemical Society, Washington, DC., pp. 115–140.

Weatherwax R.G. and Stamm A.J. (1946) *The coefficients of thermal expansion of wood and wood products*. Report No. 1487, US Forest Products Laboratory, Madison.

Chapter 5

Flow in timber

5.1 Introduction

The term *flow* is synonymous with the passage of liquids through a porous medium such as timber, but the term is also applicable to the passage of gases, thermal energy and electrical energy. It is this wider interpretation of the term that is applied in this chapter, albeit that the bulk of the chapter is devoted to the passage of both liquids and gases (i.e. fluids).

The passage of fluids through timber can occur in one of two ways, either as bulk flow through the interconnected cell lumens or other voids, or by diffusion. The latter embraces both the transfer of water vapour through air in the lumens and the movement of bound-water within the cell wall (Figure 5.1). The magnitude of the bulk flow of a fluid through timber is determined by its *permeability*.

Looking at the phenomenon of flow of moisture in wood from the point of view of the type of moisture, rather than the physical processes involved as described above, it is possible to identify the involvement of three types of moisture:

- free liquid water in the cell cavities giving rise to bulk flow above the fibre saturation point (see Section 4.2.5);
- bound water within the cell walls which moves by diffusion below the fibre saturation point (see Section 4.2.5);
- water vapour which moves by diffusion in the lumens both above and below the fibre saturation point.

It is convenient when discussing flow of any type to think of it in terms of being either constant or variable with respect to either time or location within the specimen; flow under the former conditions is referred to as *steady-state flow*, whereas when flow is time and space dependent it is referred to as *unsteady-state flow*, and these are treated separately below, although emphasis is placed on the former.

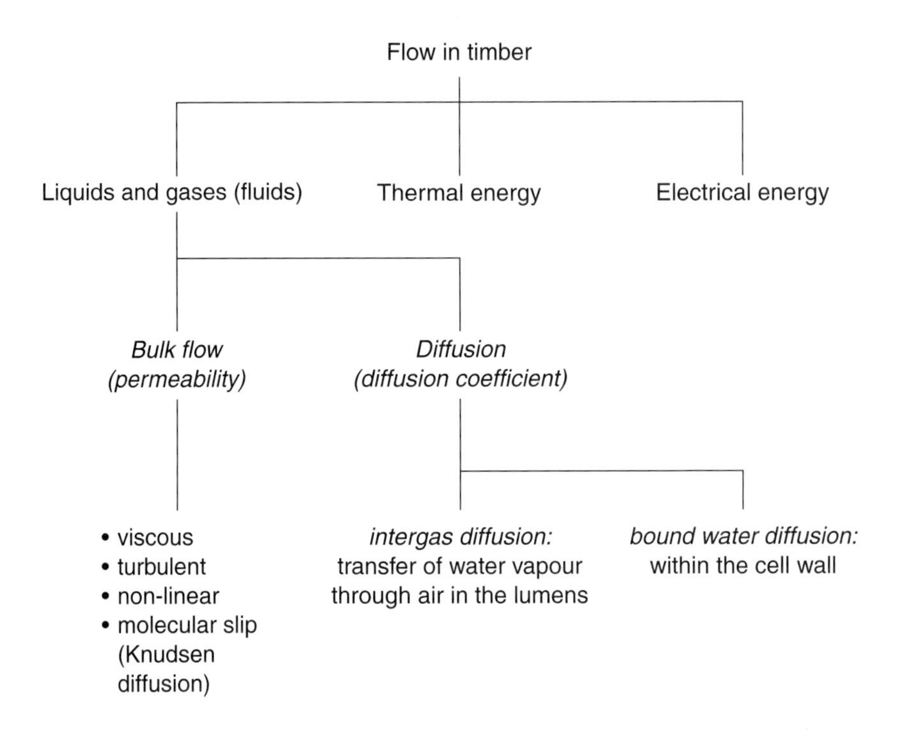

Figure 5.1 The different aspects of flow in timber which are covered in this chapter.

5.2 Steady-state flow

One of the most interesting features of flow in timber in common with many other materials is that the same basic relationship holds irrespective of whether one is concerned with liquid or gas flow, diffusion of moisture, or thermal and electrical conductivity. The basic relationship is that the flux or rate of flow is proportional to the pressure gradient:

$$\frac{\text{flux}}{\text{gradient}} = k \tag{5.1}$$

where the flux is the rate of flow per unit cross-sectional area, the gradient is the pressure difference per unit length causing flow, and k is a constant, dependent on form of flow, such as permeability, diffusion or conductivity. This relationship when relating to fluid flow is ascribed to Darcy, whereas where thermal conductivity is involved the equation is referred to as Fourier's law. However, this is of secondary importance because the significant point is that the same general relationship applies relating rate of flow to the pressure gradient irrespective of whether it is fluid flow by way of the cell cavities, moisture diffusion along the cell lumens or through the cell wall, or thermal or electrical flow along the cell walls. Each of these variants of flow will now be discussed, but greater emphasis will be placed upon fluid flow because of its practical significance.

5.2.1 Bulk flow and permeability

Permeability is simply the quantitative expression of the bulk flow of fluids through a porous material. Flow in the steady-state condition is best described in terms of Darcy's law. Thus,

$$\text{permeability} = \frac{\text{flux}}{\text{gradient}} \tag{5.2}$$

and for the flow of liquids this becomes

$$k = \frac{QL}{A\,\Delta P} \tag{5.3}$$

where k is the permeability (cm^2/(atm s)), Q is the volume rate of flow (cm^3/s), ΔP is the pressure differential (atm), A is the cross-sectional area of the specimen (cm^2), and L is the length of the specimen in the direction of flow (cm).

Because of the change of pressure of a gas and hence its volumetric flow rate as it moves through a porous medium, Darcy's law for the flow of gases has to be modified as follows:

$$k_g = \frac{QLP}{A\,\Delta P\,\bar{P}} \tag{5.4}$$

where k_g is the gas permeability and Q, L, A and ΔP are as in equation (5.3), P is the pressure at which Q is measured, and $\bar{P}$ is the mean gas pressure in the sample (Siau, 1984).

Of all the numerous physical and mechanical properties of timber, permeability is by far the most variable; when differences between timbers and differences between the principal directions within a timber are taken into consideration, the range is of the order of 10^7. Not only is permeability important in the impregnation of timber with artificial preservatives, fire retardants and stabilising chemicals, but it is also significant in the chemical removal of lignin in the manufacture of wood pulp and in the removal of *free* water during drying.

5.2.1.1 Flow of fluids

Four types of flow can occur, (see Figure 5.1) and although all four are applicable to gases, usually only the first three are relevant to the flow of liquids.

VISCOUS OR LAMINAR FLOW

Viscous or laminar flow occurs in capillaries where the rate of flow is relatively low and when the viscous forces of the fluid are overcome in shear, thereby producing an even and smooth flow pattern. In viscous flow Darcy's law is directly applicable, but a more specific relationship for flow in capillaries is given by the Poiseuille equation, which for liquids is

$$Q = \frac{N\pi r^4 \, \Delta P}{8\eta L} \tag{5.5}$$

where N is the number of uniform circular capillaries in parallel, Q is the volume rate of flow, r is the capillary radius, ΔP is the pressure drop across the capillary, L is the capillary length and η is the viscosity. For gas flow, the above equation has to be modified slightly to take into account the expansion of the gas along the pressure gradient. The amended equation is

$$Q = \frac{N\pi r^4 \, \Delta P \, \bar{P}}{8\eta L P} \tag{5.6}$$

where $\bar{P}$ is the mean gas pressure within the capillary and P is the pressure of gas where Q was measured. In both cases

$$Q \propto \frac{\Delta P}{L} \tag{5.7}$$

or flow is proportional to the pressure gradient, which conforms with the basic relationship for flow.

TURBULENT FLOW

This occurs as the rate of flow increases significantly, causing eddies to form and disrupting the smooth laminar flow; the energy required to move a given quantity of fluid is increased greatly. Turbulent flow has eluded exact mathematical description because of its complexity, though it has been shown experimentally that, approximately,

$$Q \propto \sqrt{\Delta P} \tag{5.8}$$

which means that Darcy's law is not upheld.

Turbulent flow occurs in a long cylindrical tube when the dimensionless Reynold's number exceeds approximately 2000. Reynolds number is given by

$$\frac{2dQ}{\pi r \eta} \tag{5.9}$$

where d is the density of the fluid. However, in timber it is unlikely that turbulent flow occurs, because of the relatively low fluid velocities attained and the small diameter of the capillaries present.

NON-LINEAR FLOW

At the time of the previous edition of this text, this component of flow was not recognised. However, in the late 1970s research showed that a component of flow could be ascribed to losses in kinetic energy when a moving liquid or gas enters a short capillary. This flow component was termed non-linear flow. Siau and Petty (1979) indicated that the pressure differential (ΔP) is proportional to the square of the volume rate of flow, a relationship similar to that noted above for turbulent flow. However, non-linear flow occurs at a much lower Reynolds number than is the case for turbulent flow.

Non-linear flow must be considered when short capillaries are used in the measurement of flow. In this case, readers are referred to Siau (1984) for discussion on its quantification and use.

MOLECULAR DIFFUSION

Molecular diffusion is of relevance only in gases, becoming significant when the gas pressure in a capillary is lowered to such an extent that the mean free path of the gas becomes approximately equal to the diameter of the capillary. Under these conditions the flow rate under a pressure gradient exceeds that predicted by Poiseuille's law. It is thought that this extra component of flow is the result of specular reflection occurring between the gas molecules and the capillary wall, in addition to the intermolecular collisions responsible for the laminar or viscous forces. This process can be interpreted as a sliding or slipping layer of molecules immediately adjacent to the capillary wall and is known as *molecular slip flow*.

Eventually a state is reached where further reduction of pressure results in the virtual cessation of intermolecular collisions and consequently the absence of viscous forces. Gases then progress entirely by specular reflection from the capillary walls, a process known as *Knudsen flow*. Both slip flow and Knudsen flow are diffusion processes being controlled by the concentration gradient of the molecules concerned.

In the Knudsen region, where the viscous forces are virtually absent, flow of gas in N parallel capillaries of circular section is given by

$$Q = 3N\sqrt{\frac{RT}{M}} \cdot \frac{r^3\,\Delta P}{LP} \qquad (5.10)$$

where R is the universal gas constant, T is absolute temperature and M is the molecular weight of the gas.

Slip flow occurs in wood cells predominately at pressures equal to or less than atmospheric. Gas flow with slip in timber will be given by the combination of equations (5.6) and (5.10), giving

$$Q = \frac{N\pi r^4 \, \Delta P \, \bar{P}}{8\eta LP} + 3N\sqrt{\frac{RT}{M}} \cdot \frac{r^3 \, \Delta P}{LP} \tag{5.11}$$

It follows that when slip flow occurs, the gas permeability varies according to the value of the mean pressure $\bar{P}$.

When the contribution of slip flow to total flow is significant, equation (5.11), even with the above slip term, can only be applied to homogeneous media, or to a heterogeneous collection of capillaries conducting in parallel. Darcy's law will not be upheld at low mean gas pressures for a heterogeneous porous medium composed of different types of capillary in series: the gaseous permeability of such a medium will vary non-linearly with mean gas pressure at low mean gas pressures (Petty, 1970).

Timber departs from the ideal case because capillaries are not perfect cylinders and the flow paths may involve short capillaries. As equations (5.5) and (5.6) are applicable only for long tubes, a correction factor is necessary for flow in short tubes such as occur in timber; the correction involves the increase in the length term by an amount dependent on the ratio r/L.

The theory above has been modified for flow through tubes differing from the perfect cylindrical capillaries considered above. Thus, the Kozeny–Carman equation has been applied to flow through tracheid lumina of rectangular cross-section (Petty and Puritch, 1970), whereas theory due to Hanks and Weissberg has been applied to flow through pit membrane pores which are considerably wider than they are long (Petty, 1974). In many cases, use of such theory involves considerable approximation because pores in wood are seldom entirely regular.

5.2.1.2 *Measurement of permeability*

Numerous techniques have been described over the years for the determination of permeability of timber; perhaps the gas permeability apparatus that has received most widespread approval is that developed by Petty and Preston (1969). Siau (1984) gives further details. Determination of the permeability of the cell wall material is described by Palin and Petty (1981).

5.2.1.3 *Flow paths in timber*

SOFTWOODS

Because of their simpler structure and their greater economic significance, much more attention has been paid to flow in softwood timbers than in hardwood timbers. It will be recalled from Chapter 1 that both tracheids and parenchyma cells have closed ends and that movement of liquids and gases must be by way of the pits in the cell wall. Three types of pit are present. The first is the bordered pit (Figure 1.10) which is almost entirely restricted to the radial walls of the tracheids, tending to be located towards the ends of the cells. The second type of pit is the ray or semi-bordered pit which interconnects the vertical tracheid with the horizontal ray parenchyma cell, while the third type is the simple pit between adjacent parenchyma cells.

For very many years it was firmly believed that as the diameter of the pit opening or of the openings between the margo strands was very much less than the diameter of the cell cavity, and as permeability is proportional to a power function of the capillary radius (equation (5.5)), the bordered pits would be the limiting factor controlling longitudinal flow. However, it has been demonstrated that this concept is falacious and that at least 40% of the total resistance to longitudinal flow in *Abies grandis* sapwood that had been specially dried to ensure that the torus remained in its natural position could be accounted for by the resistance of the cell cavity (Petty and Puritch, 1970).

Both longitudinal and tangential flow paths in softwoods are predominantly by way of the bordered pits as illustrated in Figure 5.2, whereas the horizontally aligned ray cells constitute the principal pathway for radial flow. It has also been suggested that very fine capillaries within the cell wall may contribute slightly to radial flow. The rates of radial flow are found to vary very widely between species.

It is not surprising to find that the different pathways to flow in the three principal axes result in anisotropy in permeability. Permeability values quoted in the literature illustrate that for most timbers longitudinal permeability is about 10^4 times the transverse permeability; mathematical modelling of longitudinal and tangential flow supports a degree of anisotropy of this order. As both longitudinal and tangential flow in softwoods are associated with bordered pits, a good correlation is to be expected between them; radial permeability is only poorly correlated with that in either of the other two directions and is frequently found to be greater than tangential permeability.

Permeability is not only directionally dependent, but is also found to vary with moisture content, between earlywood and latewood, between sapwood and heartwood (Figure 5.3) and between species. In the sapwood of green timber the torus of the bordered pit is usually located in a central position and flow can be at a maximum (Figure 5.4(a)). As the earlywood cells possess larger and more frequent bordered pits, the flow through the earlywood is considerably greater than that through the latewood. However, on drying, the torus of the earlywood cells becomes aspirated (Figure 5.4(b)) owing, it is thought, to the tension stresses set up by the retreating water meniscus (Hart and Thomas, 1967). In this process the margo strands obviously undergo very considerable extension and the torus is rigidly held in a displaced position by strong hydrogen bonding. Petty (1972) has suggested that the extension of the margo strands could be as high as 8% in the earlywood.

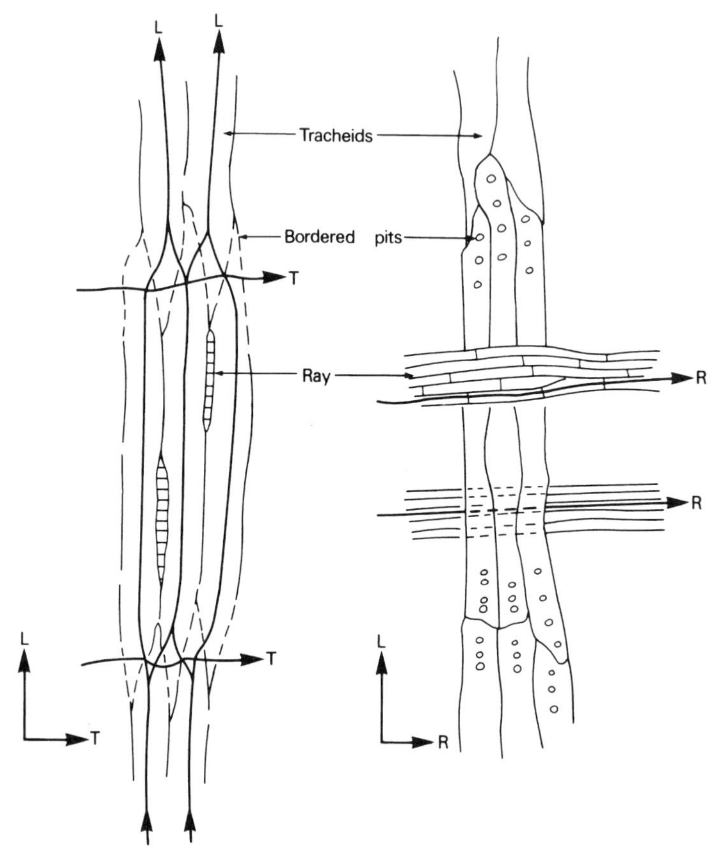

Figure 5.2 On the left is a representation of the cellular structure of a softwood in a longitudinal–tangential plane illustrating the significance of the bordered pits in both longitudinal and tangential flow. On the right is shown softwood timber in the longitudinal–radial plane, indicating the role of the ray cells in defining the principal pathway for radial flow (© BRE.)

This displacement of the torus effectively seals the pit and markedly reduces the level of permeability of dry earlywood. In the latewood, the degree of pit aspiration is very much lower on drying than in the earlywood, a phenomenon that is related to the smaller diameter and thicker cell wall of the latewood pit. Thus in dry timber, in marked contrast to green timber, the permeability of the latewood is at least as good as that of the earlywood and may even exceed it (Figure 5.3). Rewetting of the timber causes only a partial reduction in the number of aspirated pits and it appears that aspiration is mainly irreversible.

It is possible to prevent aspiration of the bordered pits by replacing the water in the timber by acetone and ethanol by solvent exchange, thus reducing the interfacial forces occurring during drying. Although this is a useful procedure

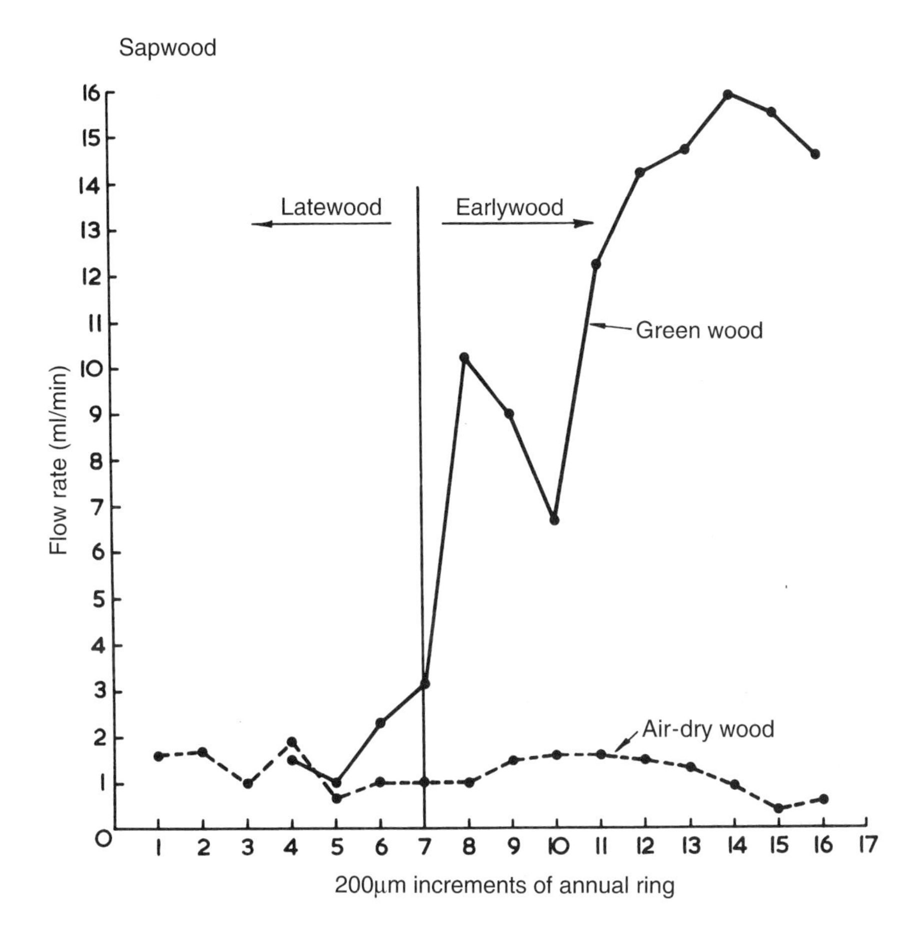

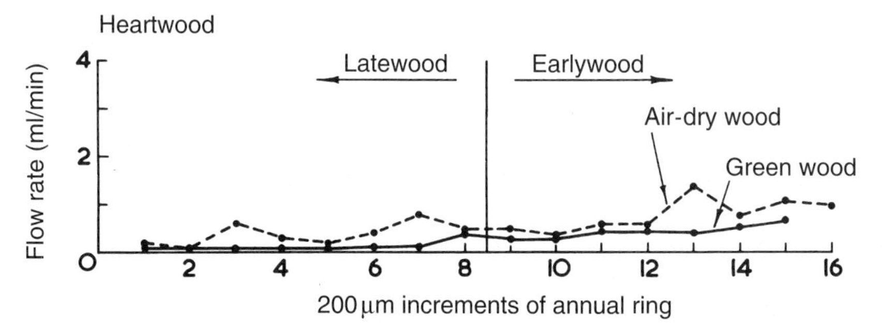

Figure 5.3 The variation in rate of longitudinal flow through samples of green and dry earlywood and latewood of Scots pine sapwood and heartwood. (From Banks (1968), © BRE.)

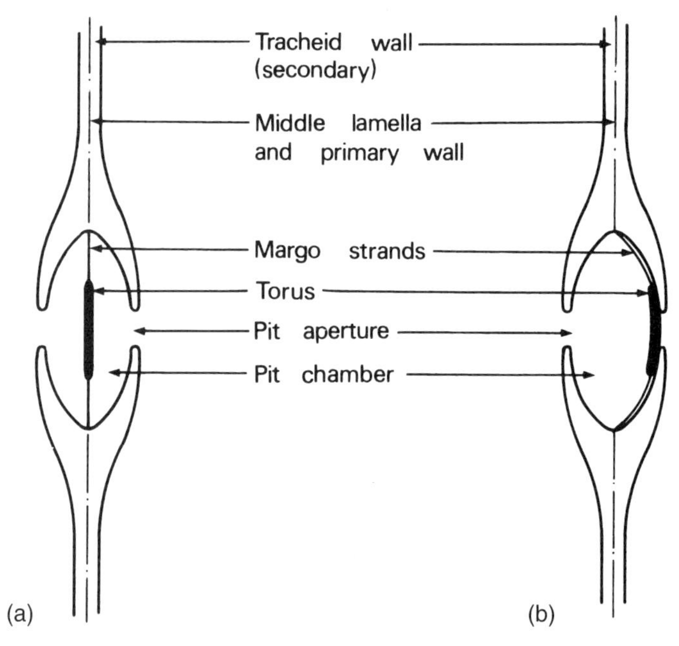

Figure 5.4 Cross-section of a bordered pit in the sapwood of a softwood timber: (a) in timber in the green condition with the torus in the 'normal' position; and (b) in timber in the dried state with the torus in an aspirated position. (© BRE.)

in research for the examination of the torus in an unaspirated state, it cannot be applied commercially because of the cost.

Quite apart from the fact that many earlywood pits are aspirated in the heartwood of softwoods, the permeability of the heartwood is usually appreciably lower than that of the sapwood due to the deposition of encrusting materials over the torus and margo strands and also within the ray cells (Figure 5.3).

Permeability varies widely among different species of softwoods. Thus, Comstock (1970) found that the ratio of longitudinal-to-tangential permeability varied between 500:1 and 80,000:1. Generally, the pines are much more permeable than the spruces, firs or Douglas fir. This can be attributed primarily, though not exclusively, to the markedly different type of semi-bordered pit present between the vertical tracheids and the ray parenchyma in the pines (fenestrate or pinoid type) compared with the spruces, firs or Douglas fir (piceoid type).

HARDWOODS

The longitudinal permeability is usually high in the sapwood of hardwoods. This is because these timbers possess vessel elements, the ends of which have been either completely or partially dissolved away. Radial flow is again by way of the rays, whereas tangential flow is more complicated, relying on the presence of pits interconnecting adjacent vessels, fibres and vertical parenchyma; however, intervascular pits in sycamore have been shown to provide considerable resistance to flow (Petty, 1981). Transverse flow rates are usually much lower than in the softwoods but, somewhat surprisingly, a good correlation exists between tangential and radial permeability; this is due in part to the very low permeability of the rays in hardwoods.

As the effects of bordered pit aspiration, so dominant in controlling the permeability of softwoods, are absent in hardwoods, the influence of drying on the level of permeability in hardwoods is very much less than is the case with softwoods.

Permeability is highest in the outer sapwood, decreasing inwards and reducing markedly with the onset of heartwood formation as the cells become blocked either by the deposition of gums or resins or, as happens in certain timbers, by the ingrowth into the vessels of cell wall material of neighbouring cells, a process known as the formation of *tyloses*.

Permeability varies widely among different species of hardwoods (Smith and Lee, 1958). This variability is due in large measure to the wide variation in vessel diameter that occurs among the hardwood species. Thus, the ring-porous hardwoods which are characterised by having earlywood vessels that are of large diameter, generally have much higher permeabilities than the diffuse-porous timbers that have vessels of considerably lower diameter; however, in those ring-porous timbers that develop tyloses (e.g. the white oaks) their heartwood permeability may be lower than that in the heartwood of diffuse-porous timbers. Interspecific variability in permeability also reflects the different types of pitting on the end walls of the vessel elements.

5.2.1.4 Timber and the laws of flow

The application of Darcy's law to the permeability of timber is based on a number of assumptions, not all of which are upheld in practice. Among the more important are that timber is a homogeneous porous material and that flow is always viscous and linear; neither of these assumptions is strictly valid, but the Darcy law remains a useful tool with which to describe flow in timber. The assumptions on which the law is based, and its limitations when applied to timber are reviewed by Scheidegger (1974).

Initial research on liquid permeability indicated that Darcy's law did not appear to be valid as the flow rate decreased with time. Further investigation has revealed that the initial results were an artefact caused by the presence of air and impurities in the impregnating liquid. By de-aeration and filtration of their liquid, many workers have since been able to achieve steady-state flow, and to find that in very general terms Darcy's Law was upheld in timber (e.g. Comstock, 1967).

A second consequence of Darcy's law is that the rate of flow is inversely related to the viscosity of the liquid. Once again the early work on flow was bedevilled by a lack of appreciation of all the factors involved and it was not until all variables occurring within the experimental period were eliminated that this second consequence was satisfied.

Gas, because of its lower viscosity and the ease with which steady flow rates can be obtained, is a most attractive fluid for permeability studies. It has been shown in previous sections that timber is a complex medium composed of different types of capillary in series, and that with such a medium deviations from Darcy's law for gases are to be expected at low mean gas pressures, due to the presence of slip flow; such deviations have been observed by a number of investigators. At higher mean gas pressures, an approximately linear relationship between conductivity and mean pressure is expected and this, too, has been observed experimentally. However, at even higher mean gas pressures, flow rate is sometimes less than proportional to the applied pressure differential due, it is thought, to the onset of non-linear flow. Darcy's law may thus appear to be valid only in the middle range of mean gas pressures.

In a study of gas flow through *Abies grandis*, Smith and Banks (1971) have verified experimentally that the contribution of laminar flow to total flow is inversely proportional to the viscosity of the gas, as is expected from equation (5.6), and that the contribution of slip flow to total flow is inversely related to the square root of the molecular weight of the gas, as is expected from equation (5.10).

The flow rate has also been found to decrease with increasing specimen length at a rate faster than that predicted by Darcy, at least up to 20 mm, but thereafter to conform with Darcy's law. This phenomenon is explained in terms of the increased probability of encountering a cell with all the pits aspirated as the specimen length is increased; therefore, the permeability of dried timber will decrease with increasing specimen length until a length is reached where all the easy pit-flow paths are blocked and flow is controlled by 'residual' flow paths whose probability tends to zero (Figure 5.5). The residual flow paths are assumed to be the unaspirated bordered pits of the latewood or, exceptionally, some of the semi-bordered ray pits (Banks, 1970; Bramhall, 1971).

Bramhall (1971), working on Douglas fir, incorporated a decay function in the Darcy equation to account for the random occurrence of aspirated pits and the reduced number of conducting tracheids with increasing depth of penetration:

$$Q = \frac{kA\,\Delta p}{L}\,\mathrm{e}^{-bL} \tag{5.12}$$

where Q is the volume flow rate, k is the permeability, A is the cross-sectional area, e is the base of natural logs, b is the positive decay constant, Δp is the pressure differential and L is the specimen length. This model, which differs

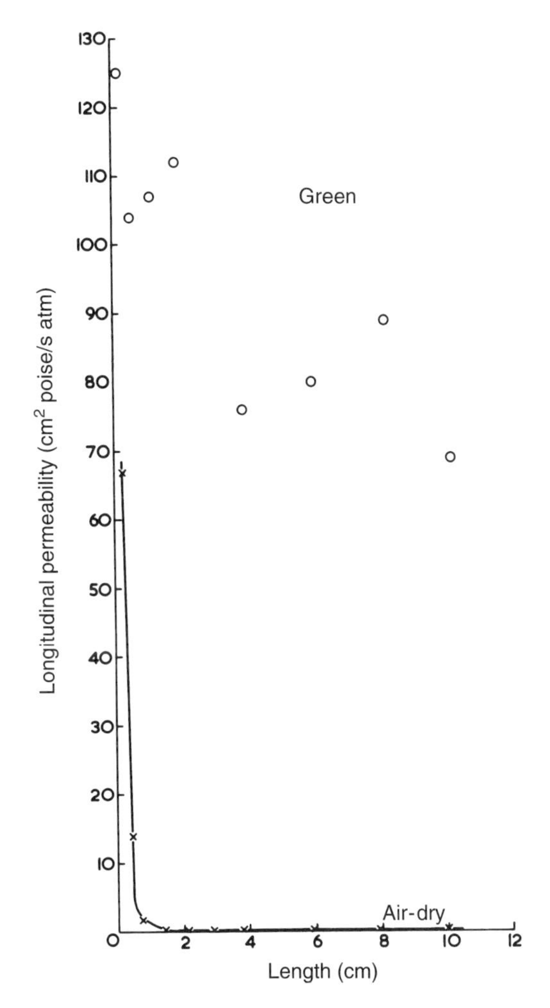

Figure 5.5 The marked decrease in longitudinal permeability of air dry samples of Norway spruce sapwood of increasing length. Permeability of the green samples with respect to length varies randomly about a mean value. (From Banks (1970), © BRE.)

from the Darcy equation by the inclusion of the exponential term, gave good agreement with experimental data.

In theory, it should be a simple matter to relate the laminar flow of gases to that of liquids through a given medium. In practice, this is difficult with a heterogeneous medium such as timber because of the presence of slip flow of the gases, and the resultant apparent departure from Darcy's law at low mean gas pressures. Experimental and analytical approaches to the solution of this problem have been moderately successful (Comstock, 1967; Smith and Banks, 1971).

To summarise, too little is known about flow in hardwoods to make any firm conclusions, but for softwoods it may be concluded that Darcy's law is only partially applicable. Certainly, the rate of flow in the laminar region is proportional to the applied pressure for both liquids and gases, but the rate of flow is not always inversely related to specimen length, nor is it proportional to pressure at the higher pressure levels. However, for gases the Darcy equation can be modified by the incorporation of a slip-flow term.

In the absence of externally applied pressure, the movement of free water is caused by capillary forces and the flow can be described in terms of the Poiseuille equation (equation (5.5)). However, as the molecules of water adjacent to the capillary walls are bound to the capillary by chemisorption, a term must be introduced in the equation to account for the disturbance of flow. Flow is produced by differences in tension due to surface forces in the menisci within the capillaries. However, the very variable dimensions of both the cell cavities and the pores in the pit membranes render almost impossible any numerical evaluation of capillary movement of moisture in timber; this is most unfortunate in view of the practical significance of moisture movement in the drying of timber.

The significance of pit aspiration in reducing capillary flow or permeability has already been discussed. One practical manifestation of the existence of capillary forces occurs when water is removed too rapidly in drying. When the capillary tension exceeds the compression strength of the timber perpendicular to the grain the cells collapse, resulting in a corrugated surface to the timber.

Unlike other properties of wood, relatively little modelling of permeability is recorded in the literature – Smith and Lee (1958), Petty (1970), Comstock (1970) and Siau *et al.* (1981).

5.2.2 *Moisture diffusion*

Flow of water below the fibre saturation point embraces both the diffusion of water vapour through the void structure comprising the cell cavities and pit membrane pores and the diffusion of bound water through the cell walls (Figure 5.1). In passing, it should be noted that because of the capillary structure of timber, vapour pressures are set up and vapour can pass through the timber both above and below the fibre saturation point; however, the flow of vapour is usually regarded as being of secondary importance to that of both bound and free water.

Moisture diffusion is another manifestation of flow, conforming with the general relationship between flux and pressure. Thus, it is possible to express diffusion of moisture in timber at a fixed temperature in terms of Fick's first law, which states that the flux of moisture diffusion is directly proportional to the gradient of moisture concentration; as such, it is analogous to the Darcy law on flow of fluids through porous media.

The total flux F of moisture diffusion through a plane surface under isothermal conditions is given by

$$F = \frac{\mathrm{d}m}{\mathrm{d}t} = -D\frac{\mathrm{d}c}{\mathrm{d}x} \tag{5.13}$$

where $\mathrm{d}m/\mathrm{d}t$ is the flux (rate of mass transfer per unit area), $\mathrm{d}c/\mathrm{d}x$ is the gradient of moisture concentration (mass per unit volume) in the x direction, and D is the moisture diffusion coefficient which is expressed in m^2/s (Siau, 1984; Skaar, 1988).

Equation (5.13) can alternatively be written in terms of percentage moisture content M as

$$F = -K_M\frac{\mathrm{d}M}{\mathrm{d}x} \tag{5.14}$$

where the transport coefficent $K_M = D \cdot \partial c/\partial M$. As the mass of water per unit volume of wood c is $\rho M/100$, it follows that $K_M = \rho D/100$, if ρ is the basic density of the wood.

Under steady-state conditions the diffusion coefficent is given by

$$D = \frac{100mL}{tA\rho\,\Delta M} \tag{5.15}$$

where m is the mass of water transported in time t, A is the cross-sectional area, L is the length of the wood sample, and ΔM is the moisture content difference driving the diffusion.

As moisture content and vapour pressure are related quantities, the latter can be considered as the driving potential for moisture diffusion, so another form of the diffusion equation is

$$F = -K_p\frac{\mathrm{d}p}{\mathrm{d}x} \tag{5.16}$$

where F is again the rate of mass transfer per unit area, p is the vapour pressure and $K_p = D \cdot \partial c/\partial p$. K_p is often referred to as the *vapour permeability*, but it should not be confused with fluid permeability k. The term $\partial c/\partial p$ is linearly related to the slope of the graph of equilibrium moisture content against relative humidity for wood (Skaar, 1988). It therefore varies in a complex manner with humidity, moisture content and temperature.

The vapour component of the total flux is usually much less than that for the bound water. The rate of diffusion of water vapour through timber at moisture contents below the fibre saturation point has been shown to yield coefficients similar to those for the diffusion of carbon dioxide, provided corrections are made for differences in molecular weight between the gases. This means that water vapour must follow the same pathway through timber as does carbon dioxide and implies that diffusion of water vapour through the cell walls is negligible in comparison to that through the cell cavities and pits (Tarkow and Stamm, 1960).

Bound water diffusion occurs when water molecules bound to their sorption sites by hydrogen bonding receive energy in excess of the bonding energy, thereby allowing them to move to new sites. At any one time the number of molecules with excess energy is proportional to the vapour pressure of the water in the timber at that moisture content and temperature. The rate of diffusion is proportional to the concentration gradient of the migrating molecules, which in turn is proportional to the vapour pressure gradient.

The most important factors affecting the diffusion coefficient of water in timber are temperature, moisture content and density of the timber. Thus, Stamm (1959) has shown that the bound water diffusion coefficient of the cell wall substance increases with temperature approximately in proportion to the increase in the saturated vapour pressure of water, and increases exponentially with increasing moisture content at constant temperature. The diffusion coefficient has also been shown to decrease with increasing density and to differ according to the method of determination at high moisture contents. It is also dependent on grain direction – the ratio of longitudinal to transverse coefficients is approximately 2.5.

Various alternative ways of expressing the potential that drives moisture through wood have been proposed. These include percentage moisture content, relative vapour pressure, osmotic pressure, chemical potential, capillary pressure (Hunter, 1995) and spreading pressure, the last mentioned being a surface phenomenon derivable from the surface sorption theory of Dent which in turn is a modification of the BET sorption theory (see Section 4.2.7; Skaar & Babiak, 1982). Although all this work has led to much debate on the correct flow potential, it has no effect on the calculation of flow. Moisture flow is the same irrespective of the potential used, provided the mathematical conversions between transport coefficients, potentials and capacity factors are carried out correctly (Skaar,1988).

Non-isothermal moisture movement has attracted little interest. However, Skaar and Siau (1981) present three equations for the diffusion of water through the cell wall under a combined moisture and temperature gradient. All three equations give similar results at low moisture contents, but the results diverge at high moisture contents.

As with the use of the Darcy equation for permeability, so with the application of Fick's law for diffusion there appears to be a number of cases in which the law is not upheld and the model fails to describe the experimental data. Claesson (1997) in describing some of the failures of Fickian models claims that this is due to a complicated, but transient sorption in the cell wall; it certainly cannot be explained by high resistance to flow of surface moisture.

The diffusion of moisture through wood has considerable practical significance because it relates to the drying of wood below the fibre saturation point, the day to day movement of wood through diurnal and seasonal changes in climate, and in the quantification of the rate of vapour transfer through a thin sheet such as the sheathing used in timber-frame construction.

Because timber and wood-based panel products are hygroscopic, the wet cup test method is commonly used to determine their vapour resistance. In this procedure the cup contains a salt solution that will produce a high humidity and a wood or panel sample of uniform thickness forms the lid of the cup; the assembly loses mass when placed in an atmosphere of low humidity. Full details of these tests are given in BS 7374, DIN 52615, ASTM C355 and, more recently, the new European standard ENISO 12572 (method C).

From the cup test, the *permeance* W is determined from

$$W = \frac{F'}{A\,\Delta p} \tag{5.17}$$

where W is in kg/(m^2 s Pa) (although in the UK this is usually reduced to g/MN s, where 1 g/MN s = 10^9 kg/(m^2 s Pa), F' is the moisture flow rate as measured by the rate of mass change of the cup assembly in kg/s, A is the area of the test piece, and Δp is the vapour pressure difference (Pa) across the test piece.

The reciprocal of vapour permeance is vapour resistance, Z, and is an expression of the vapour resistance per unit area, per unit vapour pressure differential.

Permeance and resistance relate to the specific thickness used in the wet cup test. To generalise this, the results are expressed in terms of thickness (usually 1 m) when they are known as vapour *permeability* and *resistivity* respectively. Thus, the vapour permeability is given by

$$\delta = WL \tag{5.18}$$

where δ is the permeability in kg/m s Pa (or in the UK in g m/MN s, where 1 g m/MN s = 10^9 kg/m s Pa), W is the permeance as in equation (5.17), and L is the thickness of the test piece in metres. By combining equations (5.17) and (5.18) we obtain

$$\frac{F'}{A} = \delta \frac{\Delta p}{L} \tag{5.19}$$

and comparison with equation (5.16) shows that K_p and δ are the same quantity.

Throughout much of continental Europe and also in the new European standard, moisture resistivity is expressed as the ratio of the moisture resistance of the material to the moisture resistance of air of the same thickness and temperature. This is known as the *water vapour resistance factor*, μ. Thus,

$$\mu = \frac{\delta_a}{\delta} \tag{5.20}$$

where δ_a is the water vapour permeability of air, which is the reciprocal of the moisture resistance of the air. The calculation of δ_a is given in ENISO 12572.

Examples of the water vapour resistance factor for timber and for wood-based panels are given in Table 5.1.

5.2.3 *Thermal and electrical conductivity*

The basic law for flow of thermal energy is ascribed to Fourier and when described mathematically is

$$K_h = \frac{HL}{tA \, \Delta T} \tag{5.21}$$

where K_h is the thermal conductivity for steady state flow of heat through a slab of material, H is the quantity of heat, t is time, A is the cross-sectional area, L is the length, and ΔT is the temperature differential. This equation is analogous to that of Darcy for fluid flow.

Compared with permeability, where the Darcy equation was shown to be only partially valid for timber, thermal flow is explained adequately by the Fourier equation, provided the boundary conditions are defined clearly.

Thermal conductivity will increase slightly with increased moisture content, especially when calculated on a volume-fraction-of-cell-wall basis; however, it appears that conductivity of the cell wall substance is independent of moisture content (Siau, 1984). Conductivity is influenced considerably by the density of the timber, i.e. by the volume-fraction-of-cell-wall substance, and various empirical and linear relationships between conductivity and density have been established. Conductivity will also vary with timber orientation due to its anisotropic structure: the longitudinal thermal conductivity is about 2.5 times the transverse conductivity.

Table 5.1 Generic values of the water vapour resistance factor μ for solid wood and for wood-based panel products derived using the wet cup method in ENISO 12572 (method C); ρ = density at 12% moisture content (kg/m^3)

Material	*Water vapour resistance factor* (μ)
Solid wood ρ = 300	30
Softboard	5
Particleboard ρ = 600	15
MDF	20
Cement-bonded particleboard	30
Standard hardboard	50
Plywood ρ = 700	90
OSB	130

Compared with metals, the thermal conductivity of timber is extremely low, though it is generally up to eight times higher than that of insulating materials (Table 5.2). The transverse value for timber is about one quarter that for brick, thereby explaining the lower heating requirements of timber houses compared with the traditional brick house.

Thermal insulation materials in the UK are usually rated by their U-value, where U is the conductance or the reciprocal of the thermal resistance. In North America these materials are rated in terms of their R-value or thermal resistance. Thus

$$\text{U-value} = \frac{1}{\text{R-value}} = \frac{K_h}{L} \tag{5.22}$$

where K_h is the thermal conductivity and L is the thickness of the material.

Electrical conductivity once again conforms to the basic relationship between flux (current) and pressure gradient (voltage gradient), but it is much more sensitive to moisture content than is the case with thermal conductivity; as the moisture content increases from zero to fibre saturation point the electrical conductivity increases by at least 10^{10} times. Conductivity is also temperature dependent, each 10 °C rise in temperature doubling the electrical conductivity.

Table 5.2 Thermal conductivity of timber and other materials; ρ = density in kg/m^3.

Material	K_h *(W m^{-1} K^{-1})*
Copper	400
Aluminium	201
Concrete	1.5
Glass	1.1
Brick wall	1.0
Cell-wall substance parallel to grain	0.88
Water	0.59
Cell-wall substance perpendicular to grain	0.44
Timber parallel to grain, 12% moisture content	0.38
Cement-bonded particleboard	0.23
Hardboard	0.16
Timber perpendicular to grain, 12% moisture content	0.15
Polyisoprene rubber	0.15
Plywood ρ = 500	0.13
OSB	0.13
Plaster	0.13
Particleboard ρ = 600	0.12
MDF ρ = 800	0.12
Cork (baked slab)	0.05
Glass wool	0.04
Polystyrene, cellular	0.035

Electrical conductivity, or rather its reciprocal the *resistivity*, is a most useful measure of the moisture content of timber, and mention has already been made in Chapter 4 of the use of electrical resistance meters for the rapid determination of moisture content below the fibre saturation point.

5.3 Unsteady-state flow

As recorded above, steady-state flow in timber has received considerable attention, but unfortunately, those processes that depend on the movement of fluids into or out of timber are concerned with unsteady-state flow which is usually most complex in analysis, necessitating many simplifying assumptions and consequently receiving but little attention. Unsteady-state flow occurs when the rate of flow and the pressure gradient are varying in space and time, as actually occurs during the drying of timber or its impregnation with chemical solutions. The same broad similarities in behaviour between fluids, moisture vapour, and heat that are seen with steady-state flow are again recognisable in unsteady-state flow.

5.3.1 *Unsteady-state flow of liquids*

The unsteady-state equation can be derived from the steady-state equation by differentiating with respect to time. Thus, Darcy's law for liquids may be written in derivative form (assuming parallel, circular capillaries) as

$$\frac{\mathrm{d}V}{\mathrm{d}t} = \frac{kA\,\Delta P}{x} \tag{5.23}$$

where k is the permeability, x is penetration, V is the volume of liquid and $\mathrm{d}V = \nu A\mathrm{d}x$ where ν is the porosity.

By integration it is possible to obtain both the penetration and the volumetric retention, both of which are proportional to the square root of time (Siau, 1984), thus

$$x = \sqrt{\left(\frac{2k.\Delta P.t}{\nu}\right)} \tag{5.24}$$

and

$$F_{\mathrm{VL}} = \frac{1}{L} \cdot \sqrt{\left(\frac{2k.\Delta P.t}{\nu}\right)} \tag{5.25}$$

where F_{VL} is the fraction of voids filled with liquid and L is the maximum possible penetration.

These relationships have been confirmed experimentally for several permeable softwoods by Petty (1978), who also refined equation (5.25) and added a term to account for the combined effects of capillary forces, vapour pressure of the liquid, and the pressure of the dissolved air. Good agreement between predicted and measured fractional volume was obtained.

All the studies of unsteady-state fluid flow through timber to date have assumed that timber is a homogeneous porous medium and that Darcy's law may be used as a basis for unsteady-state flow theory. However, the validity of such an approach has been questioned on the basis that the Darcian flow model is fundamentally imprecise when applied to unsteady-state flow in wood. Not only does permeability decrease with increasing sample length, but there is also the high probability that because of the wide variation in pore diameter in a sample, liquid will penetrate some flow paths more rapidly than others (Banks, 1981).

5.3.2 *Unsteady-state flow of gases*

The unsteady-state equation can be derived from the steady-state equation by differentiating with respect to time. For gaseous flow, in partial derivative form, this becomes

$$\frac{\partial P^2}{\partial t} = \bar{D}_p \frac{\partial^2 P^2}{\partial x^2} \tag{5.26}$$

where P is the gas pressure and $\bar{D}_p$ is the diffusion coefficient for hydrodynamic flow based on average pressure, averaged over space and time, which equals $k_g\bar{P}/\nu$ where k_g is the gas permeability, $\bar{P}$ is the average pressure, and ν is the porosity (Siau, 1984).

There is almost complete absence of analytical solutions for this diffusion equation when it is used to describe the flow of compressible gases. This results from the very complex variation that occurs in the diffusion coefficient with changing pressure.

Early investigations on unsteady-state flow resolved the problem by using a constant value for the diffusion coefficient. Such an assumption is not valid for gas flow because the coefficient varies systematically with pressure, which in turn is dependent on time and location. Although these early results have to be interpreted with caution, they did indicate the very marked anisotropy in the values of the diffusion coefficient with a ratio of longitudinal to transverse of over 10 000.

The problem has been resolved by the use of a dimensionless, pressure-dependent dynamic flow coefficient (equivalent to the diffusion coefficient) and the equation solved for particular boundary conditions (Sebastian et al., 1973). By replacement of the non-linear differential equation by a finite-difference equation and solution by digital computer, both the viscous and the slip flow parameters for longitudinal flow have been determined. Despite the simplified mathematical model that was adopted, the numerical solution of the partial differential equation was in reasonable agreement with the experimental data.

Alternative and mathematically more complex solutions have been derived over the years by Siau, first in 1976 and later in 1984.

5.3.3 Unsteady-state moisture diffusion

As in the case of fluid flow, the unsteady-state equation for moisture diffusion can be derived by differentiation from the steady-state equation. In the partial derivative form it becomes

$$\frac{\partial M}{\partial t} = \bar{D}\frac{\partial^2 M}{\partial x^2} \tag{5.27}$$

where M is the percentage moisture content and $\bar{D}$ is the average diffusion coefficient for moisture movement corresponding to the overall average moisture content being equal to $100\bar{K}_M/G\rho_w$ where $\bar{K}_M$ is the average conductivity coefficient for moisture diffusion corresponding to the average moisture content, $G\rho_w/100$ is the mass of moisture to raise the moisture content of a unit cube of timber by 1%, G is the specific gravity of moist wood and ρ_w is the normal density of water. This equation is referred to as Fick's second law of diffusion. The diffusion coefficient can be derived experimentally assuming that it is constant. However, once again this assumption is invalid because it is now known that the diffusion coefficient is concentration dependent. In order to circumvent the otherwise extremely complex solution of the equation, average values for the diffusion coefficient for selected moisture content ranges have been used.

5.3.4 Unsteady-state thermal conductivity

The equivalent equation in partial derivative form for thermal flow is

$$\frac{\partial T}{\partial t} = D_h\frac{\partial^2 T}{\partial x^2} \tag{5.28}$$

where D_h is the thermal diffusivity (cm^2/s) equals $K_h/c\rho$, where K_h is the thermal conductivity of gross timber (W/(m °C), and $c\rho$ is the quantity of heat required to increase the temperature of a unit cube of timber by 1°C.

Once again the diffusivity is not constant, being dependent on the thermal conductivity, density and moisture content. The ratio of the coefficients in the longitudinal and transverse directions is about 2.5.

5.4 Conclusions

The work done so far on unsteady-state flow has been informative, but caution must be exercised in any attempt to accept the findings in absolute terms because many of the assumptions on which the analysis of unsteady-state flow is based have been rendered untenable by recent findings under steady-state conditions.

Although it has been possible to obtain good correlations between theory and experimental results of both steady- and unsteady-state flow when carried out under laboratory conditions, generally rather disappointing relationships have been found under practical conditions as, for example, in the artificial seasoning of timber, the migration of moisture in buildings, or the impregnation of timber with artificial preservatives. Generally, the rates of flow obtained are very much lower than theory would indicate and unfortunately the reasons for the discrepancy are not understood. Many of the coefficients used in timber, particularly the diffusion coefficients, have been derived empirically; flow in timber is only one of many aspects of material performance where theory and practice have yet to be fully reconciled.

References

Standards and specifications

ENISO 12572 *Building materials – determination of water vapour transmission.*

Literature

Banks, W.B. (1968) A technique for measuring the lateral permeability of wood. *J. Inst. Wood Sci.*, **4** (2), 35–41.

Banks, W.B. (1970) Some factors affecting the permeability of Scots pine and Norway spruce. *J. Inst. Wood Sci.*, **5** (1), 10–17.

Banks, W.B. (1981) Addressing the problem of non-steady liquid flow in wood. *Wood Sci. Technol.*, **15**, 171–177.

Bramhall, G. (1971) The validity of Darcy's Law in the axial penetration of wood, *Wood Sci. Technol.*, **5**, 121–134.

Claesson, J. (1997) Mathematical modelling of moisture transport. *Proc. International conference on wood-water relations,* ed. P Hoffmeyer, Copenhagen, and published by the management committee of EC COST Action E8, 61–68.

Comstock, G.L. (1967) Longitudinal permeability of wood to gases and nonswelling liquids. *For. Prod. J.*, **17** (10), 41–46.

Comstock, G.L. (1970) Directional permeability of softwoods. *Wood & Fiber*, **1**, 283–289.

Hart, C.A. and Thomas, R.J. (1967) Mechanism of bordered pit aspiration as caused by capillarity. *For. Prod. J.*, **17** (11), 61–68.

Hunter, A.J. (1995) Equilibrium moisture content and the movement of water through wood above fibre saturation. *Wood Sci. Technol.*, **29**, 129–135.

Palin, M.A. and Petty, J.A. (1981) Permeability to water of the cell wall material of spruce heartwood. *Wood Sci. Technol.*, **15**, 161–169.

Petty, J.A. (1970) Permeability and structure of the wood of Sitka spruce. *Proc. Roy. Soc. Lond.*, **B175**, 149–166.

Petty, J.A. (1972) The aspiration of bordered pits in conifer wood. *Proc. Roy. Soc. Lond.*, **B181**, 395–406.

Petty, J.A. (1974) Laminar flow of fluids through short capillaries in conifer wood. *Wood Sci. Technol.*, **8** (4), 275–282.

Petty, J.A. (1978) Influence of viscosity and pressure on the initial absorption of non-swelling liquids by pine sapwood. *Holzforschung*, **32**, 134–137.

Petty, J.A. (1981) Fluid flow through the vessels and intervascular pits of sycamore wood. *Holzforschung*, **35**, 213–216.

Petty, J.A. and Preston R.D. (1969) The dimensions and number of pit membrane pores in conifer wood. *Proc. Roy. Soc. Lond.*, **B172**, 137–151.

Petty, J.A. and Puritch, G.S. (1970) The effects of drying on the structure and permeability of the wood of *Abies grandis*. *Wood Sci. Technol.*, **4** (2), 140–154.

Scheidegger, A.E. (1974) *The physics of flow through porous media*, 3rd edn, University of Toronto Press, Toronto.

Sebastian, L.P., Siau, J.F. and Skaar, C. (1973) Unsteady-state axial flow of gas in wood. *Wood Science*, **6** (2), 167–174.

Siau, J.F. (1976) A model for unsteady-state gas flow in the longitudinal direction in wood. *Wood Sci. Technol.*, **10**, 149–155.

Siau, J.F. (1984) *Transport processes in wood*, Springer-Verlag, Berlin.

Siau, J.F., Kanagawa, Y. and Petty, J.A. (1981) The use of permeability and capillary theory to characterise the structure of wood and membrane filters. *Wood and Fiber*, **13**, 2–12.

Siau, J.F. and Petty, J.A. (1979) Corrections for capillaries used in permeability measurements of wood. *Wood Sci. Technol.*, **13**, 179–185.

Skaar, C. (1988) *Wood-water relations*, Springer-Verlag, Berlin.

Skaar, C. and Babiak, M. (1982) A model for bound-water transport in wood. *Wood Sci. Technol.*, **16**, 123–138.

Skaar, C. and Siau, J.F. (1981) Thermal diffusion of bound water in wood. *Wood Sci. Technol.*, **15**, 105–112.

Smith, D.N.R. and Banks, W.B. (1971) The mechanism of flow of gases through coniferous wood, *Proc. Roy. Soc. Lond.*, **B177**, 197–223.

Smith, D.N.R. and Lee, E. (1958) *The longitudinal permeability of some hardwoods and softwoods*, Department of Scientific and Industrial Research, Special Report 13, London.

Stamm, A.J. (1959) Bound water diffusion into wood in the fiber direction, *For. Prod. J.*, **9** (1), 27–32.

Tarkow, H. and Stamm, A.J. (1960) Diffusion through the air filled capillaries of softwoods: I, Carbon dioxide; II, Water vapour, *For. Prod. J.*, **10**, 247–250; 323–324.

Chapter 6

Deformation under load

6.1 Introduction

This chapter is concerned with the type and magnitude of the deformation that results from the application of an external load. As in the case of both concrete and polymers, the load–deformation relationship in timber is exceedingly complex, resulting from these facts:

- Timber does not behave in a truly elastic mode, rather its behaviour is time dependent.
- The magnitude of the strain is influenced by a wide range of factors. Some of these are property dependent, such as density of the timber, angle of the grain relative to direction of load application and angle of the microfibrils within the cell wall. Others are environmentally dependent, such as temperature and relative humidity.

Under service conditions timber often has to withstand an imposed load for many years, perhaps even centuries; this is particularly relevant in construction applications. When loaded, timber will deform and a generalised interpretation of the variation of deformation with time together with the various components of this deformation is illustrated in Figure 6.1. On the application of a load at time zero an instantaneous (and reversible) deformation occurs which represents true elastic behaviour. On maintaining the load to time t_1 the deformation increases, though the rate of increase is continually decreasing; this increase in deformation with time is termed *creep*. On removal of the load at time t_1 an instantaneous reduction in deformation occurs which is approximately equal in magnitude to the initial elastic deformation. With time, the remaining deformation will decrease at an ever-decreasing rate until at time t_2 no further reduction occurs. The creep that has occurred during loading can be conveniently subdivided into a *reversible* component, which disappears with time and which can be regarded as *delayed elastic* behaviour, and an *irreversible* component which results from *plastic* or *viscous* flow. Therefore, timber on loading possesses three forms of deformation behaviour – elastic, delayed elastic and viscous. Like so many other materials, timber can be treated neither as a truly

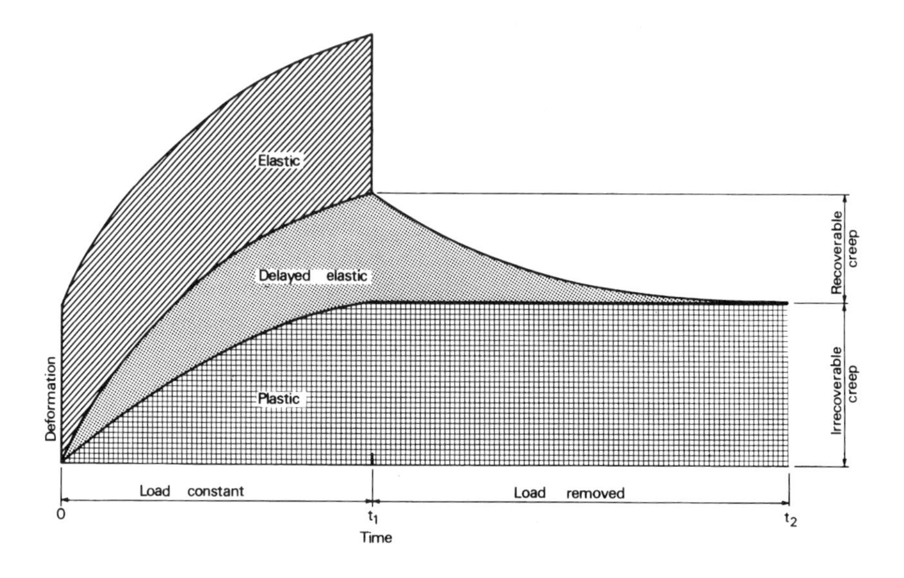

Figure 6.1 The various elastic and plastic components of the deformation of timber under constant load. (© BRE.)

elastic material where, by Hooke's law, stress (see Section 6.2) is proportional to strain but independent of the rate of strain, nor as a truly viscous liquid where, according to Newton's law, stress is proportional to rate of strain, but independent of strain itself. Where combinations of behaviour are encountered the material is said to be viscoelastic and timber, like many high polymers, is a viscoelastic material.

Having defined timber as such, the reader will no doubt be surprised to find that half of this chapter is devoted to the elastic behaviour of timber. It has already been discussed how part of the deformation can be described as elastic and the section below indicates how at low levels of stressing and short periods of time there is considerable justification for treating the material as such. Perhaps the greatest incentive for this viewpoint is the fact that classical elasticity theory is well established and, when applied to timber, has been shown to work very well. The question of time in any stress analysis can be accommodated by the use of safety factors in design calculations.

Consequently, Section 6.2 deals first with elastic deformation as representing a very good approximation of what happens in practice, then Section 6.3 deals with viscoelastic deformation, which embraces both delayed elastic and irreversible deformation. Although technically more applicable to timber, viscoelasticity is certainly less well understood and developed in its application than is the case with elasticity theory.

6.2 Elastic deformation

When a sample of timber is loaded in tension, compression or bending, the instantaneous deformations obtained with increasing load are approximately proportional to the values of the applied load. Figure 6.2 illustrates that this approximation is certainly truer of the experimental evidence in longitudinal tensile loading than in the case of longitudinal compression. In both modes of loading, the approximation appears to become a reality at the lower levels of loading. Thus, it has become convenient to recognise a point of inflection on the load–deflection curve known as the *limit of proportionality*, below which the relationship between load and deformation is linear, and above which non-linearity occurs. Generally, the limit of proportionality in longitudinal tension is found to occur at about 60% of the ultimate load to failure, whereas in longitudinal compression the limit is considerably lower, varying from 30% to 50% of the failure value.

At the lower levels of loading, therefore, where the straight-line relationship appears to be valid, the material is said to be linearly elastic. Hence

$$\text{deformation} \propto \text{applied load}$$

$$\text{i.e.} \quad \frac{\text{applied load}}{\text{deformation}} = \text{a constant}$$

The applied load must be quantified in terms of the cross-sectional area carrying that load, whereas the deflection or extension must be related to the original dimension of the test piece prior to load application. Hence

$$\frac{\text{load (N)}}{\text{cross-sectional area (mm}^2\text{)}} = \text{stress (N/mm}^2\text{), denoted by } \sigma$$

and

$$\frac{\text{deformation (mm)}}{\text{original length (mm)}} = \text{strain (unitless), denoted by } \varepsilon$$

Therefore,

$$\frac{\text{stress } (\sigma)}{\text{strain } (\varepsilon)} = \text{a constant} = E \text{ (N/mm}^2\text{)} \qquad (6.1)$$

where ε is the strain (change in dimension/original dimension), σ is the stress (load/cross-sectional area), and E is a constant known as the modulus of elasticity. The modulus of elasticity (MOE), is also referred to in the literature as the elastic modulus, Young's modulus, or simply and frequently, though incorrectly, as stiffness.

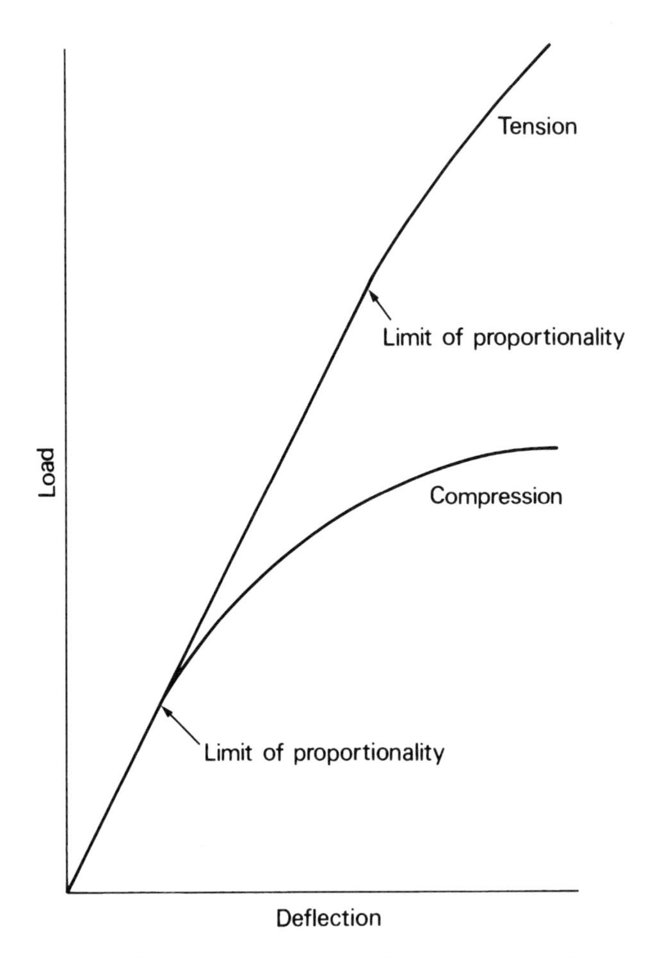

Figure 6.2 Load–deflection graphs for timber stressed in tension and compression parallel to the grain. The assumed *limit of proportionality* for each graph is indicated. (© BRE.)

The apparent linearity at the lower levels of loading is really an artefact introduced by the rate of testing. At fast rates of loading, a very good approximation to a straight line occurs but, as the rate of loading decreases, the load–deflection line assumes a curvilinear shape (Figure 6.3). Such curves can be treated as linear by introducing a straight line approximation, which can take the form of either a tangent or secant. Traditionally for timber and wood fibre composites, tangent lines have been used as linear approximations of load–deflection curves.

Thus, although in theory it should be possible to obtain a true elastic response, in practice this is rarely the case, albeit the degree of divergence is frequently very low. It should be appreciated in passing that a curvilinear load–deflection

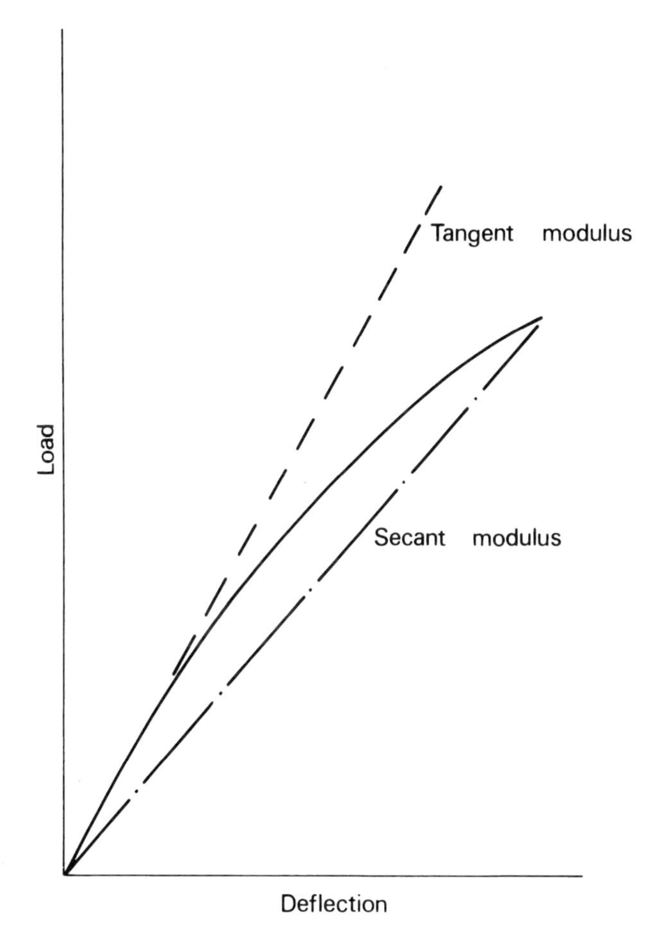

Figure 6.3 The approximation of a curvilinear load–deflection curve for timber stressed at low loading rates, by linear tangents or secants. (© BRE).

curve must not be interpreted as an absence of true elastic behaviour. The material may still behave elastically, though not linearly elastically. The prime criterion for elastic behaviour is that the load–deflection curve is truly reversible, i.e. no permanent deformation occurs on release of the load.

6.2.1 Modulus of elasticity

The modulus of elasticity or elastic modulus in the longitudinal direction is one of the principal elastic constants of the material. Although the following test methods can be modified to measure elasticity in other planes, they are described specifically for the determination of elasticity in the longitudinal direction.

6.2.1.1 Size of test piece

The size of the test piece to be used will be determined by the type of information required. When this has been decided, it will determine the test procedure; standardised test procedures should always be adopted and a choice is available between National, European, or International methods of test (see Figure 6.4).

USE OF SMALL, CLEAR TEST PIECES

In the early days this size of test piece was used for the derivation of working stresses for timber. However, since the mid-1970s this size of test piece has been superseded by structural-size timber. However, the small clear test piece still remains valid for characterising new timbers and for the strict academic comparison of wood from different trees or from different species, as the use of small knot-free straight-grained perfect test pieces represent the maximum quality of wood that can be obtained. As such, these test pieces are not representative of structural-size timber with all its imperfections. Several arbitrarily defined reduction factors have to be used in order to obtain a measure of the working stresses of the timber when small clear test pieces are used (see Sections 7.2.1.1 and 7.8.2.1).

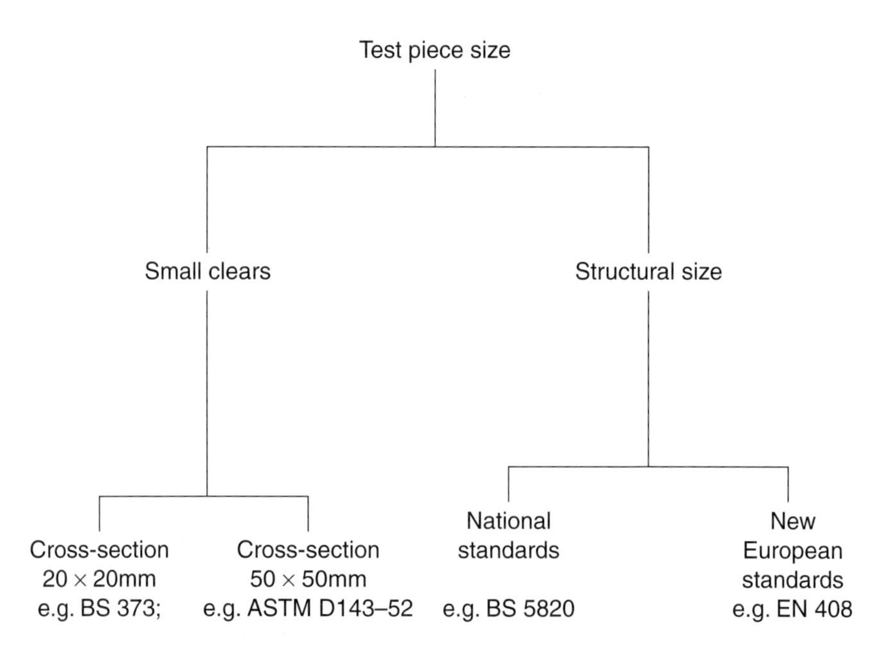

Figure 6.4 Alternative sizes of test pieces in the determination of modulus of elasticity.

USE OF STRUCTURAL-SIZE TEST PIECES

The use of these larger test pieces reproduce actual service loading conditions and they are of particular value because they allow directly for defects such as knots, splits and distorted grain, rather than by applying a series of reduction factors as is necessary with small, clear test pieces. However, use of the large pieces is probably more costly.

6.2.1.2. *Standardised test procedures*

Europe at the present time (1999) is in a transition period in which National test procedures are being replaced by European standards and specifications. Within the first decade of the twenty-first century all national standards relating to the use of timber and panel products in construction will be withdrawn. It is interesting to note that many of the European standards (ENs) have now been adopted as International Standards (ISOs).

6.2.1.3 *Test methods*

A whole spectrum of test methods, many of them fairly simple in concept, exists for the determination of the elastic modulus. These methods can be conveniently subdivided into two groups, the first comprising *static* methods based on the application of a direct stress and the measurement of the resultant strain, with the second group comprising *dynamic* methods based on resonant vibration from flexural, torsional or ultrasonic pulse excitation. It is appropriate to examine a few of these techniques.

The determination of the elastic modulus from stress–strain curves has already been described and, despite its sensitivity to rate of loading, it remains one of the more common methods. Although frequently carried out in the bending mode, it can also be derived from compression or tension tests. The value of the modulus in tensile, compressive and bending modes is approximately equal.

Where the tensile mode is adopted using small clear test pieces to BS 373 (1986), waisted samples of timber must be used because of the very high longitudinal tensile strength. Deformation under load is measured by an extensometer fitted to the waisted region and can be clearly seen in Figure 6.5. The slope of the line on the load–deflection graph, corrected for cross-sectional area of the sample and distance between the extensometer grips, provides the elastic modulus.

Where structural–size test pieces are tested in tension according to EN 408 (1995), the full cross-section is utilised and the length of timber in each grip must be at least nine times the larger of the two cross-sectional directions in order to reduce the risk of pull-out from the grips.

In the three-point bending test using small clear test pieces to BS 373 (1986), measurement of deflection by gauge or extensometer will provide the static modulus using the equation

$$E_s = \frac{Fl^3}{48I\Delta} \qquad (6.2)$$

where E_s is the static modulus of elasticity in three-point bending in N/mm^2, F is the load applied to the centre of the span at the limit of proportionality in newtons, l is the distance between the supports in millimetres, I is the second moment of area of the section determined from its actual dimensions in $(\text{millimetres})^4$, and Δ is the deflection at the centre of the span at the limit of proportionality.

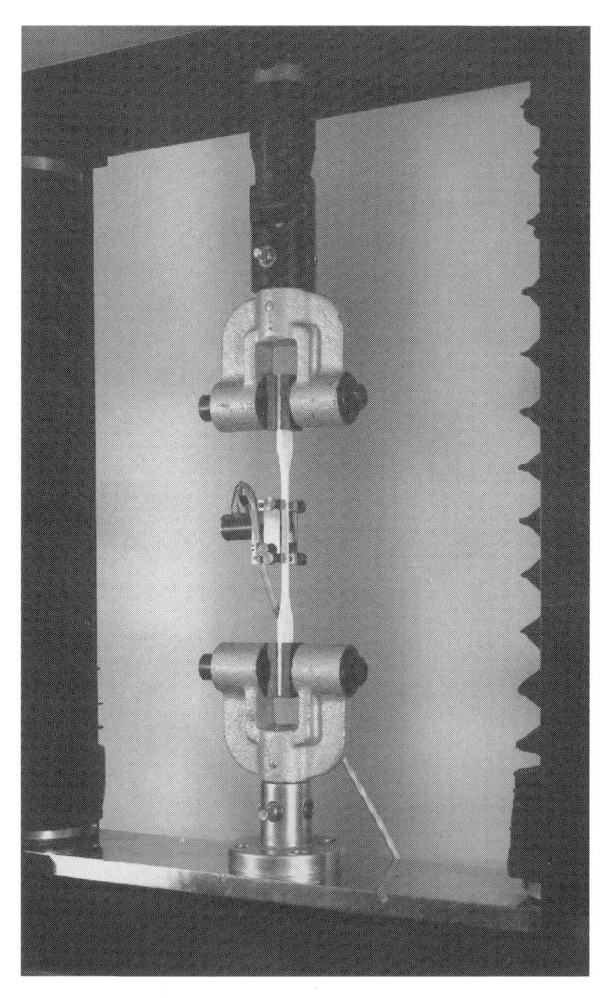

Figure 6.5 Measurement of elongation in the waisted region of a timber sample under tensile loading, in which changes in voltage across the transducer attached to the two pairs of grips provide a measure of the extension. (© BRE.)

This three-point bending modulus includes a contribution from shear deflection, whereas testing in four-point bending provides a true modulus of bending devoid of shear between the two loading points.

When the four-point bend test is carried out according to either BS 5820 (1979), or EN 408 (1995) the modulus of elasticity is calculated from the equation

$$E_m = \frac{\Delta F \, a l_1^{\,2}}{16 I \, \Delta W} \tag{6.3}$$

where E_m is the modulus of elasticity in four-point bending in newtons per square millimetre, ΔF is an increment of load over the elastic part of the load-deflection curve in newtons, a is the distance between an inner load point and the nearest support in millimetres, l_1 is the gauge length in millimetres, I is the second moment of area of the section determined from its actual dimensions in (millimetres)^4, and ΔW is the deflection under the increment of load ΔF in millimetres.

Perhaps the simplest method of determining the static modulus E_s is by loading a cantilever beam and then measuring deflection for each increment of load using, for example, a cathetometer. The static elastic modulus is then given by

$$E_s = \frac{F l^3}{3 I \Delta} \tag{6.4}$$

where F (the load applied) and Δ (deflection) are at the free end, and l is the length of the cantilever.

The determination of the dynamic elastic modulus, E_d, can be obtained by either longitudinal or flexural vibration. The latter is used more frequently and may take the form of a small unloaded beam to which are attached thin metal plates. The beam vibrates under the action of an oscillating electromagnetic impulse. The response is measured as a function of the frequency, and the dynamic elastic modulus is calculated from the specimen dimensions and resonant frequency.

The dynamic elastic modulus can be obtained very simply and accurately by clamping a strip of timber firmly to a bench in such a way that one end of the strip extends horizontally beyond the bench. A suitable mass is then attached to the free end of the strip and the period of the resulting vibrations determined (Hearmon, 1966). The dynamic modulus of elasticity is then given by

$$E_d = \frac{4\pi^2 l^3}{3 I T^2}\left(m_t + \frac{33}{140} m\right) \tag{6.5}$$

where l is the free length of the strip, I is the moment of inertia about the neutral axis ($bd^3/12$ for a rectangular strip), T is the time of period, m_t is the mass of the free length of timber, and m is the load applied.

Ultrasonic techniques can also be employed to measure E_d. These are based on the delay time of a set of pulses sent out by one transducer and received by a second; elasticity is proportional to the velocity of propagation, which will vary with grain direction.

The values of modulus obtained dynamically are generally only marginally greater than those obtained by static methods. The relationship obtained between E_d and E_s in one experiment using the same cantilever strips prepared from a number of species is presented in Figure 6.6. Although the mean value of E_d is only about 3% higher than that for E_s, the differences between the readings are nevertheless significant at the 0.1% level. Thus for timber, like concrete, it would appear that the value of the elastic modulus is dependent on test method, though the difference appears smaller for timber than for concrete.

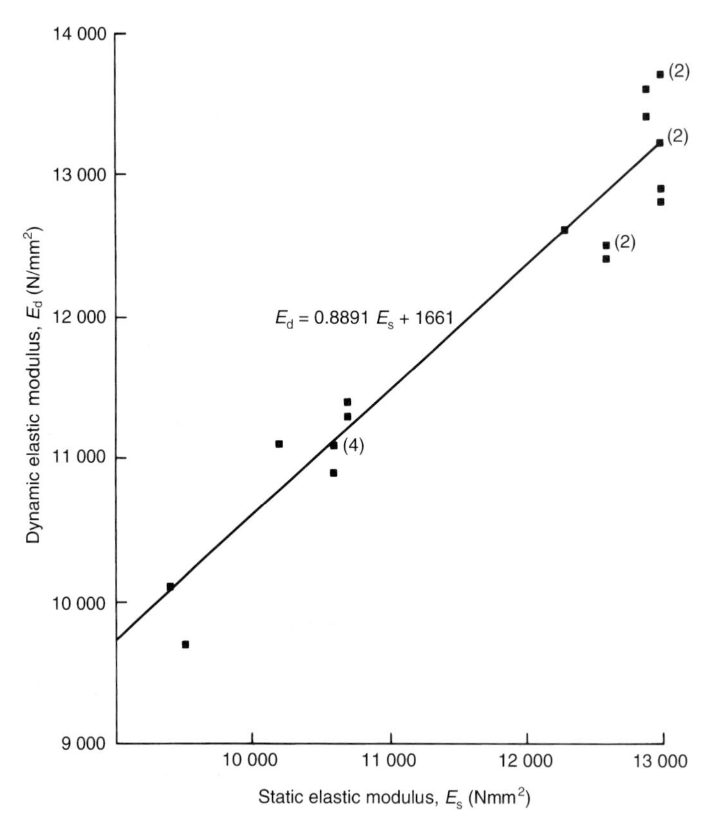

Figure 6.6 The relationship between the modulus of elasticity obtained by dynamic and static methods on the same timber samples. (© BRE.)

6.2.2 Modulus of rigidity

Within the elastic range of the material, shearing stress is proportional to shearing strain. The constant relating these parameters is called the shear or rigidity modulus and is denoted by the letter G. Thus

$$G = \frac{\tau}{\gamma} \tag{6.6}$$

where τ is the shearing stress and γ is the shearing strain.

For structural size test pieces, the modulus of rigidity (the shear modulus) can be derived by one of two test procedures as described in EN 408 (1995). In the first technique, the modulus is calculated from the difference in elastic modulus when derived by both four-point and three-point bending *on the same test piece*. In the second technique, the same test piece is loaded in three-point bending under at least four different spans.

6.2.3 Poisson's ratio

In general, when a body is subjected to a stress in one direction, the body will undergo a change in dimensions at right angles to the direction of stressing. The ratio of the contraction or extension to the applied strain is known as Poisson's ratio, and for isotropic bodies is given as

$$\nu = -\frac{\varepsilon_y}{\varepsilon_x} \tag{6.7}$$

where ε_x and ε_y are strains in the x and y directions resulting from an applied stress in the x direction. (The minus sign indicates that, when ε_x is a tensile positive strain, ε_y is a compressive negative strain.) In timber, because of its anisotropic behaviour and its treatment as a rhombic system, six Poisson's ratios occur.

6.2.4 Orthotropic elasticity and timber

The theory of elasticity is a well developed area of applied mathematics which is usually discussed in relation to isotropic materials, However, by suitable development it can be applied to materials possessing orthotropic symmetry. Providing the assumption is made that timber has orthotropic symmetry (and the justification for this assumption will be discussed later) then this form of elasticity can be applied to timber.

Let us start with the generalised stress condition and work towards the particular. If we imagine a cube of material with its edges lying along the coordinate axes, there will be a set of three mutually perpendicular stresses acting on each face. Stresses are labelled by the notation σ_{ij} where i refers to the direction of stress and j to the direction of the perpendicular to the face on which it acts (Figure 6.7). The shear stresses can be readily identified as those with $i \neq j$.

Now it is important to state that our cube will not rotate and for this to happen $\sigma_{ij} = \sigma_{ji}$, whereupon the number of components working on the cube is reduced from nine to six, three normal and three shear. Similarly it can be argued that there will be six strain components.

As the handling of these double suffixes proves awkward at times it is customary to use only single suffixes. The conversions are set out in matrix form in equations (6.8).

$$\begin{bmatrix} \sigma_{11} & \sigma_{12} & \sigma_{13} \\ - & \sigma_{22} & \sigma_{23} \\ - & - & \sigma_{33} \end{bmatrix} = \begin{bmatrix} \sigma_{1} & \sigma_{6} & \sigma_{5} \\ - & \sigma_{2} & \sigma_{4} \\ - & - & \sigma_{3} \end{bmatrix} \tag{6.8a}$$

$$\begin{bmatrix} \varepsilon_{11} & 2\varepsilon_{12} & 2\varepsilon_{13} \\ - & \varepsilon_{22} & 2\varepsilon_{23} \\ - & - & \varepsilon_{33} \end{bmatrix} = \begin{bmatrix} \varepsilon_{1} & \varepsilon_{6} & \varepsilon_{5} \\ - & \varepsilon_{2} & \varepsilon_{4} \\ - & - & \varepsilon_{3} \end{bmatrix} \tag{6.8b}$$

Note the factor 2 entered into the conversion of the shear strains. This does not affect the form of the law but proves to be mathematically convenient.

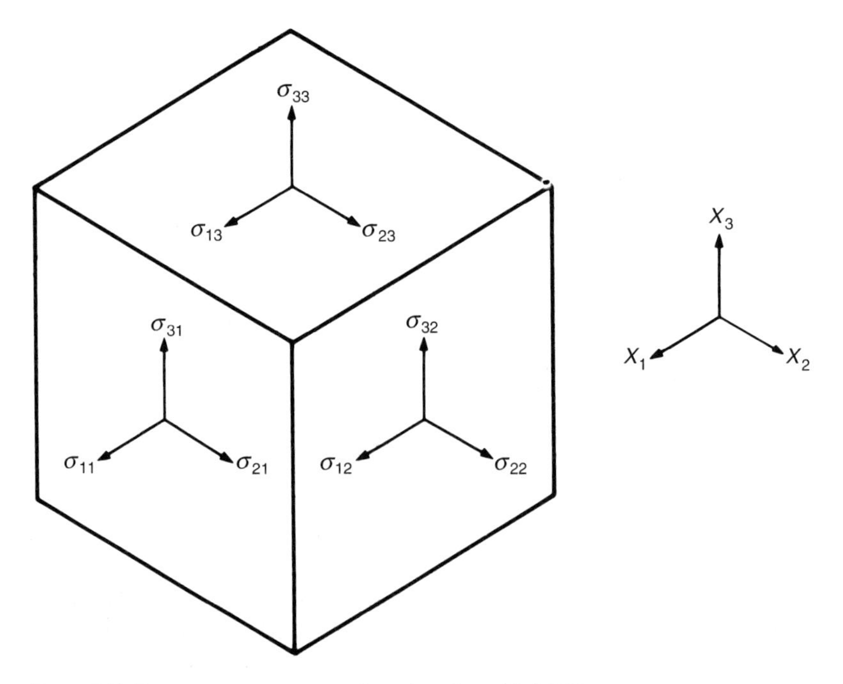

Figure 6.7 Stresses acting on a cube of timber. (© BRE.)

The relationship between stress and strain is expressed by the generalised Hooke's law, which states that each stress component is linearly related to all the strain components and *vice versa*. Now these six components of strain can be expressed in terms of the six products of the appropriate components of stress and elastic compliances (denoted by S), thus

$$\begin{aligned}
\varepsilon_1 &= S_{11}\sigma_1 + S_{12}\sigma_2 + S_{13}\sigma_3 + S_{14}\sigma_4 + S_{15}\sigma_5 + S_{16}\sigma_6 \\
\varepsilon_2 &= S_{21}\sigma_1 + S_{22}\sigma_2 + S_{23}\sigma_3 + S_{24}\sigma_4 + S_{25}\sigma_5 + S_{26}\sigma_6 \\
\varepsilon_3 &= S_{31}\sigma_1 + S_{32}\sigma_2 + S_{33}\sigma_3 + S_{34}\sigma_4 + S_{35}\sigma_5 + S_{36}\sigma_6 \\
\varepsilon_4 &= S_{41}\sigma_1 + S_{42}\sigma_2 + S_{43}\sigma_3 + S_{44}\sigma_4 + S_{45}\sigma_5 + S_{46}\sigma_6 \\
\varepsilon_5 &= S_{51}\sigma_1 + S_{52}\sigma_2 + S_{53}\sigma_3 + S_{54}\sigma_4 + S_{55}\sigma_5 + S_{56}\sigma_6 \\
\varepsilon_6 &= S_{61}\sigma_1 + S_{62}\sigma_2 + S_{63}\sigma_3 + S_{64}\sigma_4 + S_{65}\sigma_5 + S_{66}\sigma_6
\end{aligned} \tag{6.9}$$

Similarly, the stresses could be expressed in terms of all the strain components and the appropriate elastic stiffnesses. As there are six stress and six strain components, there are altogether 36 compliances (S) and 36 stiffnesses (C). However, due to thermodynamic considerations,

$$S_{ij} = S_{ji} \text{ and } C_{ij} = C_{ji} \tag{6.10}$$

which reduces the number of compliances and stiffnesses each to 21 in the most general case. However, many materials possess a particular structure such that the number of independent constants is reduced. This is the case for orthotropic materials, which possess three mutually perpendicular planes of elastic symmetry and, most important, the principal axes are chosen to be in the directions perpendicular to the orthotropic planes.

The generalised Hooke's law for orthotropic materials becomes (in matrix form)

$$\begin{bmatrix} \varepsilon_1 \\ \varepsilon_2 \\ \varepsilon_3 \\ \varepsilon_4 \\ \varepsilon_5 \\ \varepsilon_6 \end{bmatrix} = \begin{bmatrix} S_{11} & S_{12} & S_{13} & 0 & 0 & 0 \\ S_{21} & S_{22} & S_{23} & 0 & 0 & 0 \\ S_{31} & S_{32} & S_{33} & 0 & 0 & 0 \\ 0 & 0 & 0 & S_{44} & 0 & 0 \\ 0 & 0 & 0 & 0 & S_{55} & 0 \\ 0 & 0 & 0 & 0 & 0 & S_{66} \end{bmatrix} \begin{bmatrix} \sigma_1 \\ \sigma_2 \\ \sigma_3 \\ \sigma_4 \\ \sigma_5 \\ \sigma_6 \end{bmatrix} \tag{6.11}$$

The compliance matrix is thus symmetrical, comprising nine independent compliance parameters. Now an orthotropic material is characterised by six elastic moduli, three of which are the ratios of normal stress to strain in the principal directions (E) and three are the ratio of shear stress to strain in the orthotropic planes (G). The relationship of these constants to the compliances listed above can be determined by taking each in turn and assuming that only a single stress is acting. By applying stress σ_i and measuring ε_i, the slope of the stress–strain diagram can be obtained. Thus

$$\sigma_1/\varepsilon_1 = E_1;\ \sigma_2/\varepsilon_2 = E_2;\ \sigma_3/\varepsilon_3 = E_3;\ \sigma_4/\varepsilon_4 = G_{44};\ \sigma_5/\varepsilon_5 = G_{55};\ \sigma_6/\varepsilon_6 = G_{66} \quad (6.12)$$

where E is the modulus of elasticity, and G is the modulus of rigidity.

Now under these conditions

$$\varepsilon_1 = \sigma_1 S_{11};\quad \varepsilon_2 = \sigma_1 S_{21};\quad \varepsilon_3 = \sigma_1 S_{31} \quad (6.13)$$

Substituting in equation (6.12)

$$E_1 = 1/S_{11} \quad (6.14)$$

Similarly, it can be shown that

$$E_2 = 1/S_{22};\quad E_3 = 1/S_{33};\quad G_{44} = 1/S_{44};\quad G_{55} = 1/S_{55};\quad G_{66} = 1/S_{66} \quad (6.15)$$

As previously discussed, orthotropic materials are characterised by possessing six Poisson's ratios. However, three are linked with the modulus of elasticity and only three are truly independent.

Returning to the condition of the application of only a single stress, this will result in two induced strains and consequently two Poisson's ratios, thus:

$$\varepsilon_2/\varepsilon_1 = -\nu_{21} \quad \text{and} \quad \varepsilon_3/\varepsilon_1 = -\nu_{31} \quad (6.16)$$

Substituting in equation (6.13) gives

$$-\nu_{21} = S_{21}/S_{11} \quad \text{and} \quad -\nu_{31} = S_{31}/S_{11} \quad (6.17)$$

Similarly it can be shown that

$$-\nu_{12} = S_{12}/S_{22};\quad -\nu_{32} = S_{32}/S_{22}$$
$$-\nu_{13} = S_{13}/S_{33};\quad -\nu_{23} = S_{23}/S_{33} \quad (6.18)$$

Substituting in equations (6.14) and (6.15) gives

$$S_{12} = -\nu_{12}/E_2;\quad S_{13} = -\nu_{13}/E_3;\quad S_{23} = -\nu_{23}/E_3$$
$$S_{21} = -\nu_{21}/E_1;\quad S_{31} = -\nu_{31}/E_1;\quad S_{32} = -\nu_{32}/E_2 \quad (6.19)$$

Using these definitions, the constitutive equation for timber as an assumed orthotropic elastic body becomes

$$\begin{bmatrix} \varepsilon_1 \\ \varepsilon_2 \\ \varepsilon_3 \\ \varepsilon_4 \\ \varepsilon_5 \\ \varepsilon_6 \end{bmatrix} = \begin{bmatrix} 1/E_1 & -\nu_{12}/E_2 & -\nu_{13}/E_3 & 0 & 0 & 0 \\ -\nu_{21}/E_1 & 1/E_2 & -\nu_{23}/E_3 & 0 & 0 & 0 \\ -\nu_{31}/E_1 & -\nu_{32}/E_2 & 1/E_3 & 0 & 0 & 0 \\ 0 & 0 & 0 & 1/G_{44} & 0 & 0 \\ 0 & 0 & 0 & 0 & 1/G_{55} & 0 \\ 0 & 0 & 0 & 0 & 0 & 1/G_{66} \end{bmatrix} \begin{bmatrix} \sigma_1 \\ \sigma_2 \\ \sigma_3 \\ \sigma_4 \\ \sigma_5 \\ \sigma_6 \end{bmatrix} \quad (6.20)$$

Reconverting the suffixes (equation (6.8)) and substituting letters (where L, T, and R are, respectively, the longitudinal, tangential and radial directions, and the double subscripts refer to the various planes) in place of numbers, Equation (6.20) becomes

$$\begin{bmatrix} \varepsilon_{LL} \\ \varepsilon_{TT} \\ \varepsilon_{RR} \\ 2\varepsilon_{TR} \\ 2\varepsilon_{LR} \\ 2\varepsilon_{LT} \end{bmatrix} = \begin{bmatrix} 1/E_L & -\nu_{LT}/E_T & -\nu_{LR}/E_R & 0 & 0 & 0 \\ -\nu_{TL}/E_L & 1/E_T & -\nu_{TR}/E_R & 0 & 0 & 0 \\ -\nu_{RL}/E_L & -\nu_{RT}/E_T & 1/E_R & 0 & 0 & 0 \\ 0 & 0 & 0 & 1/G_{TR} & 0 & 0 \\ 0 & 0 & 0 & 0 & 1/G_{LR} & 0 \\ 0 & 0 & 0 & 0 & 0 & 1/G_{LT} \end{bmatrix} \begin{bmatrix} \sigma_{LL} \\ \sigma_{TT} \\ \sigma_{RR} \\ \sigma_{TR} \\ \sigma_{LR} \\ \sigma_{LT} \end{bmatrix}$$

Orthotropic elasticity is applied only infrequently to the solution of three-dimensional problems, owing to the complexity of the solutions. Wherever possible two-dimensional approximations are made (known as plane systems), thereby reducing the number of constitutive equations with commensurate simplification of the solution. Plane stress and plane strain systems will nearly always result in three-dimensional states of strain and stress, respectively. For example, taking the case of a plane stress situation (in the 1–2 plane) where the stress components σ_1, σ_2 and σ_6 are present and σ_3, σ_4 and σ_5 are zero, equation (6.11) will yield strain components ε_1, ε_2, ε_3 and ε_6. Now ε_3, the strain normal to the plane of the applied stress, is very small and can usually be safely ignored. Under these conditions the constitutive equation in matrix form can be written as

$$\begin{bmatrix} \varepsilon_1 \\ \varepsilon_2 \\ \varepsilon_6 \end{bmatrix} = \begin{bmatrix} S_{11} & S_{12} & 0 \\ S_{21} & S_{22} & 0 \\ 0 & 0 & S_{66} \end{bmatrix} \begin{bmatrix} \sigma_1 \\ \sigma_2 \\ \sigma_6 \end{bmatrix} \quad (6.22)$$

An important property of anisotropic materials is the change in the values of stiffness and compliance with orientation. The effect of rotation on these parameters can be obtained by transforming the stresses or strains from the rotated to the principal axes on the grounds that the potential energy of the deformed body is independent of orientation. However, the compliance tensor can be transformed directly because, by definition, it is an invariant. For an orthotropic material, transformation gives the new compliances in terms of the original compliances and the direction cosines.

In applying the elements of orthotropic elasticity to timber, the assumption is made that the three principal elasticity directions coincide with the longitudinal, radial and tangential directions in the tree. The assumption implies that the tangential faces are straight and not curved, and that the radial faces are parallel and not diverging. However, by dealing with small pieces of timber removed at some distance from the centre of the tree, the approximation of rhombic symmetry for a system possessing circular symmetry becomes more and more acceptable.

The nine independent constants required to specify the elastic behaviour of timber are: the three moduli of elasticity, one in each of the L, R and T (longitudinal, radial and tangential, respectively) directions; the three moduli of rigidity, one in each of the principal planes LT, LR and TR; and three Poisson's ratios, namely ν_{RT}, ν_{LR}, ν_{TL}. These constants, together with the three dependent Poisson's ratios ν_{RL}, ν_{TR} and ν_{LT}, are presented in Table 6.1 for a selection of hardwoods and softwoods. The table illustrates the high degree of anisotropy present in timber. Comparison of E_L with either E_R or E_T, and G_{TR} with G_{LT} or G_{RL} will indicate a degree of anisotropy, which can be as high as 60:1; usually, the ratio of E_L to E_{HORIZ} is of the order of 40:1. Note should be taken that the values of ν_{TR} are frequently greater than 0.5. (Further values of E_L are to be found in Table 7.1.)

6.2.5 *Factors influencing the elastic modulus*

The stiffness of timber is influenced by many factors, some of them properties of the material while others are components of the environment.

6.2.5.1 *Grain angle*

The significance of grain angle in influencing stiffness has already been described in the section dealing with elasticity theory. It will be recalled that the effect of rotation on the elastic constants can be obtained by transforming the stresses or strains from the rotated to the principal axes. Figure 6.8, in addition to illustrating the marked influence of grain angle on stiffness, shows the degree of fit between experimentally derived values and those obtained from transformation equations (see equation (7.11)).

Table 6.1 Values of the elastic constants for five hardwoods and four softwoods determined on small clear specimens

Species	*Density (kg/m^3)*	*Moisture content (%)*	E_L	E_R	E_T	ν_{TR}	ν_{LR}	ν_{RT}	ν_{LT}	ν_{RL}	ν_{TL}	G_{LT}	G_{LR}	G_{TR}
Hardwoods														
Balsa	200	9	6 300	300	106	0.66	0.018	0.24	0.009	0.23	0.49	203	312	33
Khaya	440	11	10 200	1130	510	0.60	0.033	0.26	0.032	0.30	0.64	600	900	210
Walnut	590	11	11 200	1190	630	0.72	0.052	0.37	0.036	0.49	0.63	700	960	230
Birch	620	9	16 300	1110	620	0.78	0.034	0.38	0.018	0.49	0.43	910	1180	190
Ash	670	9	15 800	1510	800	0.71	0.051	0.36	0.030	0.46	0.51	890	1340	270
Beech	750	11	13 700	2240	1140	0.75	0.073	0.36	0.044	0.45	0.51	1060	1610	460
Softwoods														
Norway spruce	390	12	10 700	710	430	0.51	0.030	0.31	0.025	0.38	0.51	620	500	23
Sitka spruce	390	12	11 600	900	500	0.43	0.029	0.25	0.020	0.37	0.47	720	750	39
Scots pine	550	10	16 300	1100	570	0.68	0.038	0.31	0.015	0.42	0.51	680	1160	66
Douglas fir[a]	590	9	16 400	1300	900	0.63	0.028	0.40	0.024	0.43	0.37	910	1180	79

Source: Hearmon (1948), but with different notation for the Poisson's ratios.

[a] Listed in the original as Oregon pine.
E is the modulus of elasticity in a direction indicated by the subscript (N/mm^2).
G is the modulus of rigidity in a plane indicated by the subscript (N/mm^2).
ν_{ij} is the Poisson's ratio for an extensional stress in the j direction, given by

$$\nu_{ij} = \frac{\text{compressive strain in the } i \text{ direction}}{\text{extensional strain in the } j \text{ direction}}$$

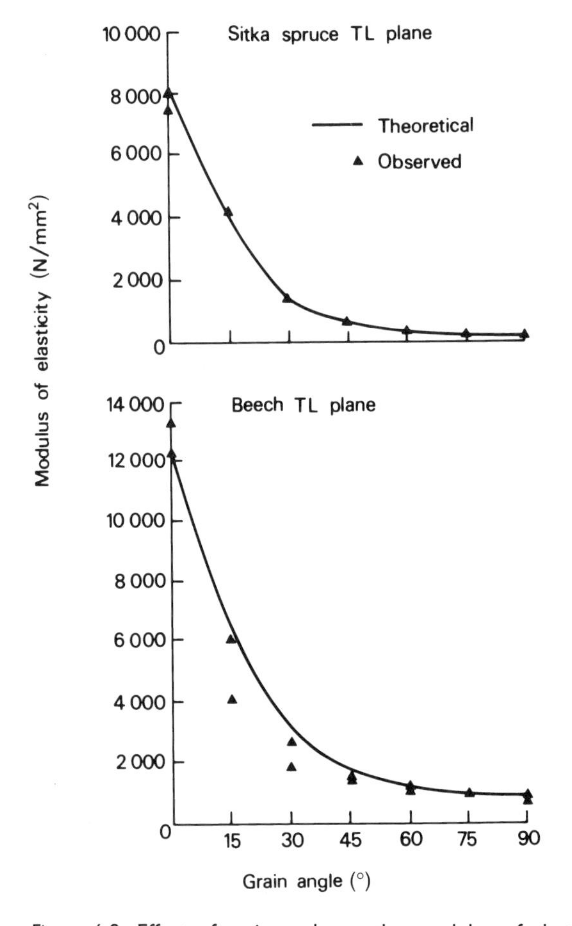

Figure 6.8 Effect of grain angle on the modulus of elasticity. (© BRE.)

6.2.5.2 Density

Stiffness is related to density of the timber, a relationship which was apparent in Table 6.1 and which is confirmed by the plot of over 200 species of timber contained in Bulletin 50 of the former Forest Products Research Laboratory (Figure 6.9). The correlation coefficient was 0.88 for timber at 12% moisture content and 0.81 for green timber and the relationship is curvilinear. A high correlation is to be expected, because density is a function of the ratio of cell wall thickness to cell diameter; consequently, increasing density will result in increasing stiffness of the cell.

Owing to the variability in structure that exists between different timbers, the relationship between density and stiffness will be higher where only a single

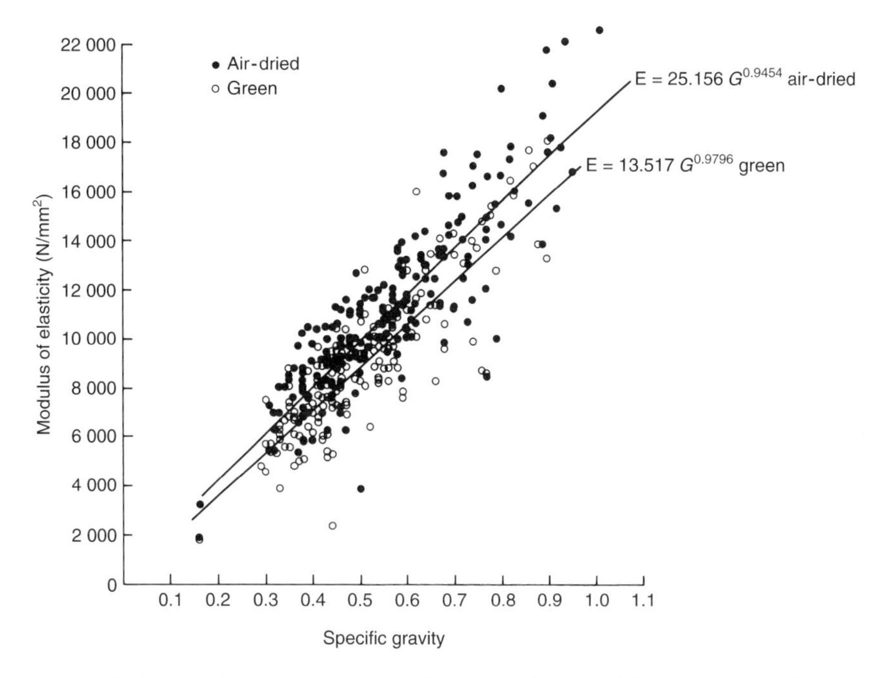

Figure 6.9 Effect of specific gravity on the longitudinal modulus of elasticity for over 200 species of timber tested in the green and dry states. (© BRE.)

species is under investigation. Because of the reduced range in density, a linear regression is usually fitted.

Similar relationships with density have been recorded for the modulus of rigidity in certain species. In others, however, for example spruce, both the longitudinal–tangential and longitudinal–radial shear moduli have been found to be independent of density. Most investigators agree that the Poisson's ratios are independent of density.

6.2.5.3 Knots

As timber is anisotropic in behaviour and knots are characterised by the occurrence of distorted grain, it is not surprising to find that the presence of knots in timber results in a reduction in the stiffness. The relationship is difficult to quantify because the effect of the knots will depend not only on their number and size, but also on their distribution both along the length of the sample and across the faces. Dead knots, especially where the knot has fallen out, will result in larger reductions in stiffness than will green knots (see Chapter 1).

6.2.5.4 Ultrastructure

Two components of the fine or chemical structure have a profound influence on both the elastic and rigidity moduli. The first relates to the existence of a matrix material with particular emphasis on the presence of lignin. In those plants devoid of lignin, such as the grasses, or in wood fibres that have been delignified, the stiffness of the cells is low and it would appear that lignin, apart from its hydrophilic protective role for the cellulosic crystallites, is responsible to a considerable extent for the high stiffness found in timber.

The significance of lignin in determining stiffness is not to imply that the cellulose fraction plays no part; on the contrary, it has been shown that the angle at which the microfibrils are lying in the middle layer of the secondary cell wall, S_2, also plays a significant role in controlling stiffness (Figure 6.10; Cowdrey and Preston, 1966; Cave, 1968).

A considerable number of mathematical models have been devised over the years to relate stiffness to the microfibrillar angle. The early models were

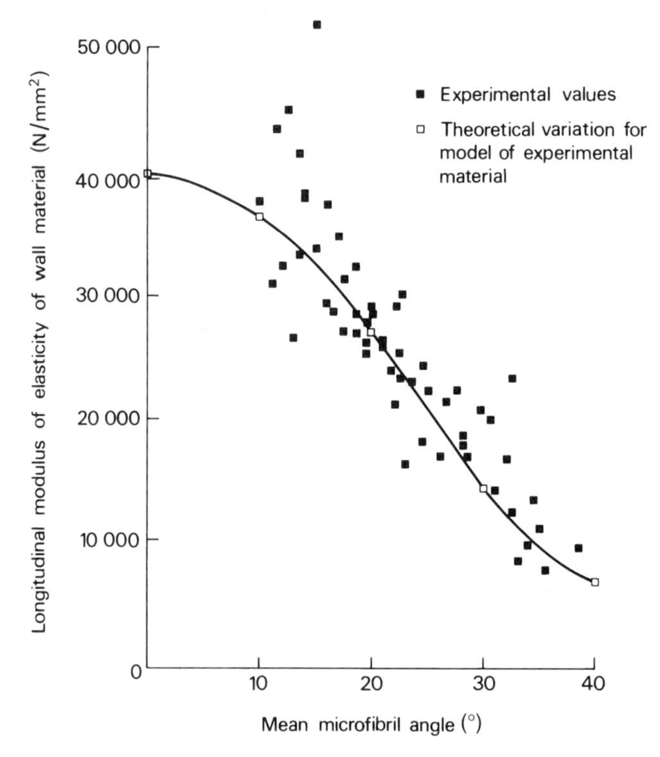

Figure 6.10 Effect of the mean microfibrillar angle of the cell wall on the longitudinal modulus of elasticity of the wall material in *Pinus radiata*. Calculated values from a mathematical model are also included. (From I. D. Cave (1968) *Wood Sci. Technol.*, **2**, 268–278, reproduced by permission of Springer-Verlag.)

two-dimensional in approach, treating the cell wall as a planar slab of material, but later the models became much more sophisticated, taking into account the existence of cell wall layers other than the S_2, the variation in microfibrillar angle between the radial and tangential walls and consequently the probability that they undergo different strains and, lastly, the possibility of complete shear restraint within the cell wall. These three-dimensional models are frequently analysed using finite element techniques. These early models were reviewed by Dinwoodie (1975).

Later modelling of timber behaviour in terms of its structure has been reviewed by Astley *et al.* (1998); it makes reference to the work of Cave (1975) who used the concept of an elastic fibre composite consisting of an inert fibre phase embedded in a water-reactive matrix. The constitutive equation is related to the overall stiffness of the composite, the volume fraction, and the stiffness and sorption characteristics of the matrix. Unlike previous models, the equation can be applied not only to elasticity, but also to shrinkage and even moisture-induced creep (Cave, 1975).

Recently, Harrington *et al.* (1998) have developed a model of the wood cell wall based on the homogenisation first, of an amorphous lignin–polyose matrix, and then of a representative volume element comprising a cellulose microfibril, its polyose-cellulose sheath and the surrounding matrix. The model predicts orthotropic elastic constants in good agreement with recorded values (Astley *et al.*, 1998). The predicted variation of axial stiffness with the S_2 microfibrillar angle is consistent with observed behaviour and aligned with results from other cell-wall models.

Stiffness of a material is very dependent on the type and degree of chemical bonding within its structure. The abundance of covalent bonding in the longitudinal plane and hydrogen bonding in one of the transverse planes contributes considerably to the moderately high levels of stiffness characteristic of timber.

6.2.5.5 Moisture content

The influence of moisture content on stiffness is similar though not quite so sensitive as that for strength which was briefly described in Chapter 4 and illustrated in Figure 4.1. Early experiments by Carrington (1922), in which stiffness was measured on a specimen of Sitka spruce as it took up moisture from the dry state, clearly indicated a linear loss in stiffness as the moisture content increased to about 30%, corresponding to the fibre saturation point as discussed in Chapter 4. Further increase in moisture content has no influence in stiffness (Figure 6.11). It will be noted from this figure that the longitudinal elastic modulus is less sensitive to changes in moisture content than either the radial or tangential elastic moduli.

Carrington also measured the rigidity moduli and Poisson's ratios. Although the former showed a similar trend to the elastic moduli, only one of the latter displayed the same trend, whereas the three other Poisson's ratios that were measured showed an inverted relationship (Figure 6.11).

Carrington's results for the variation in longitudinal moduli have been confirmed using the simple dynamic methods described earlier in this chapter. Measurement of the frequency of vibration was carried out at regular intervals as samples of Sitka spruce were dried from 70% to zero moisture content (Figure 6.12).

Confirmation of the reduction in modulus of elasticity with increasing moisture content is forthcoming, first from Figure 6.9, in which the regression lines of elasticity against density for over 200 species of timber at 12% moisture

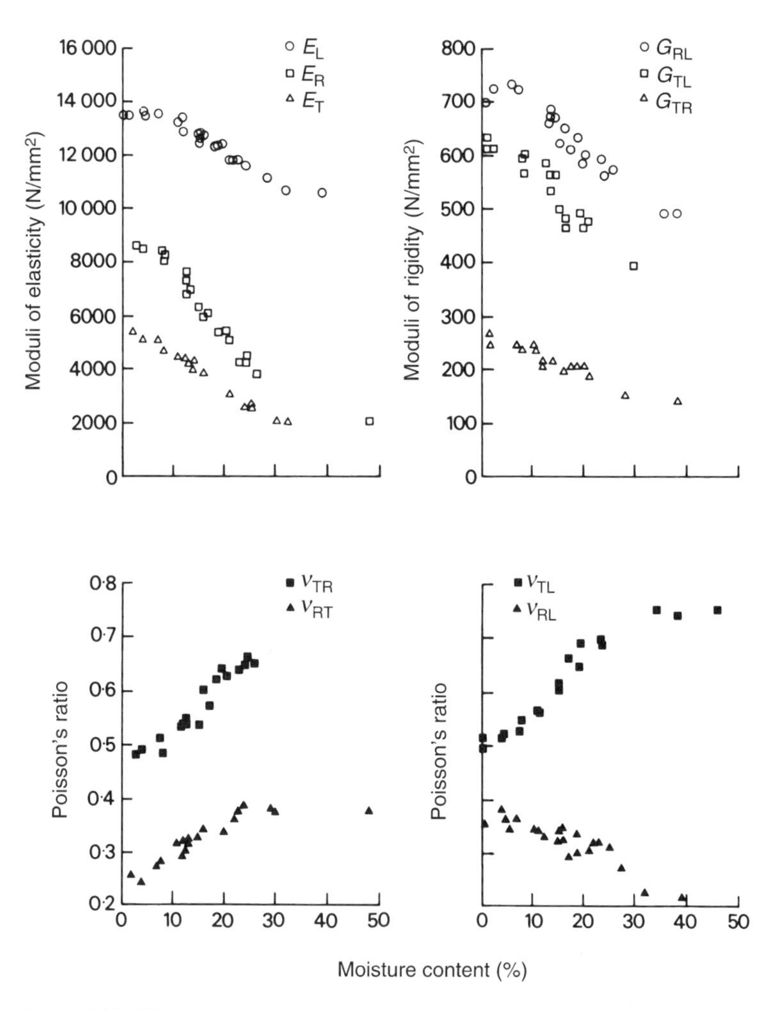

Figure 6.11 Effect of moisture content on the elastic constants of Sitka spruce. (From H. Carrington (1922) *Aeronautical Journal*, **24**, 462.)

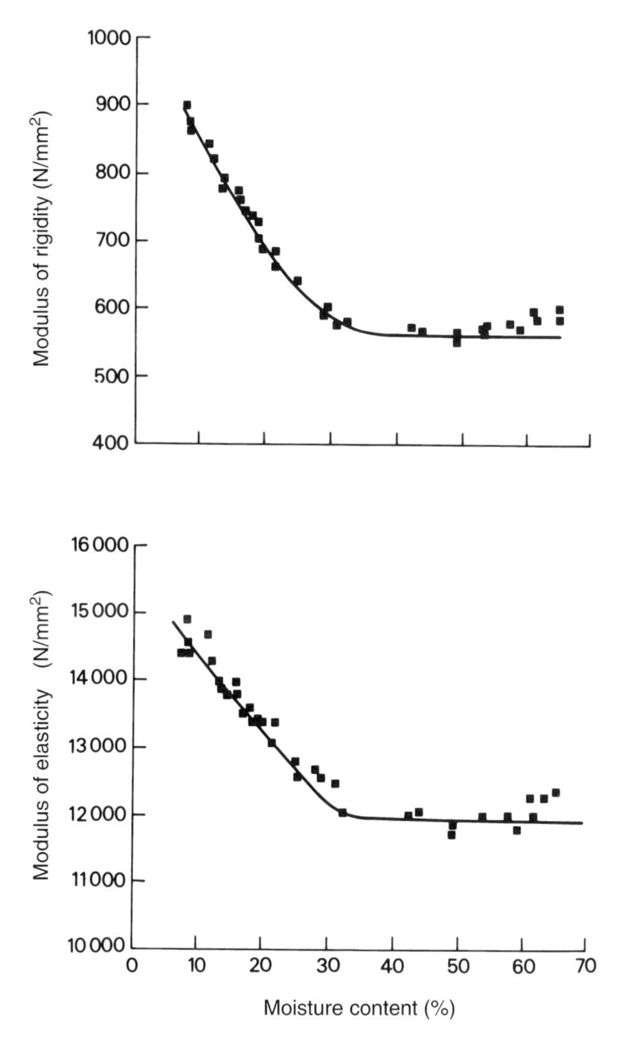

Figure 6.12 Effect of moisture content on the longitudinal modulus of elasticity and the modulus of rigidity in the LR plane in Sitka spruce. Both moduli were determined dynamically. (© BRE.)

content and in the green state are presented, and second, from the review of more recent results by Gerhards (1982).

Barkas (1945) has pointed out that when timber is stressed in compression at constant relative humidity it will give up water to the atmosphere, and conversely under tensile stressing it will absorb moisture. The equilibrium strain will therefore be the sum of that produced elastically and that caused by moisture loss or gain. It is therefore necessary to distinguish between the elastic constants at constant humidity (E_h) and those measured at constant moisture content (E_m). Hearmon (1948) has indicated that the ratio of E_h/E_m is 0.92 in the tangential direction, 0.95 in the radial direction and 1.0 in the longitudinal direction when spruce is stressed at 90% relative humidity. At 40% the ratio increases to 0.98 and 0.99 in the tangential and radial directions, respectively.

The effect of moisture increase has a far greater effect on the modulus perpendicular to the grain than on the modulus along the grain (Gerhards, 1982).

6.2.5.6 Temperature

In timber, like most other materials, increasing temperature results in greater oscillatory movement of the molecules and an enlargement of the crystal lattice. These in turn affect the mechanical properties and the stiffness and strength of the material decreases.

Although the relationship between stiffness and temperature has been shown experimentally to be curvilinear, the degree of curvature is usually slight at low moisture contents (Figure 6.13) and the relation is frequently treated as linear:

$$E_T = E_t [1 - a(T - t)] \qquad (6.23)$$

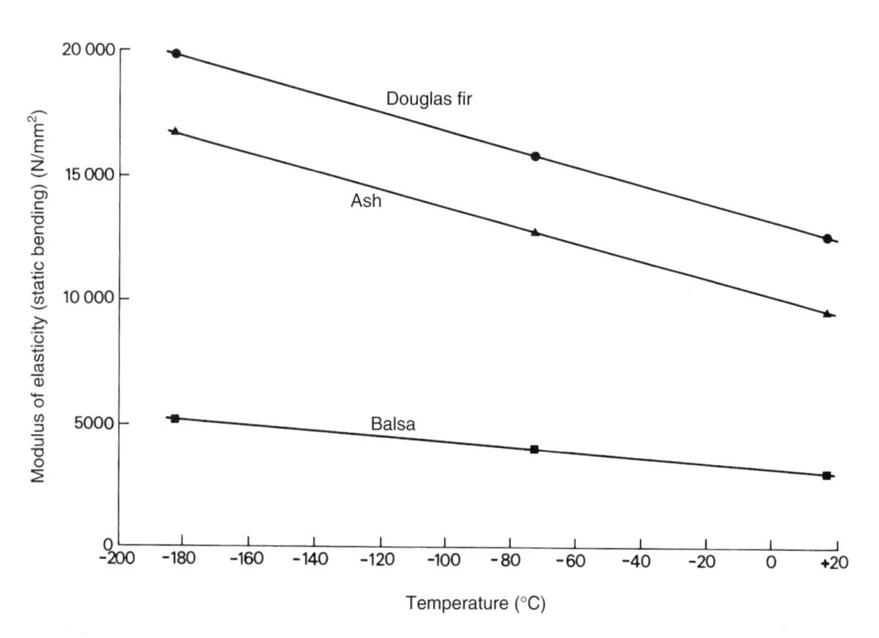

Figure 6.13 Effect of temperature on the modulus of elasticity of three species of timber. (© BRE.)

where E is the elastic modulus, T is a higher temperature, t is a lower temperature, and a is the temperature coefficient. The value a for longitudinal modulus has been shown to lie between 0.001 and 0.007 for low moisture contents.

The effect of a temperature increase is greater on the perpendicular modulus than on the longitudinal modulus (Gerhards,1982).

At higher moisture contents the relationship between stiffness and temperature is markedly curvilinear and the interaction of moisture content and temperature in influencing stiffness is clearly shown in Figure 6.14, which summarises the extensive work of Sulzberger (1947). At zero moisture content the reduction in stiffness between –20 °C and +60 °C is only 6%. At 20% moisture content the corresponding reduction is 40%. This increase in the significance of temperature with increasing moisture content has been confirmed by Gerhards (1982).

Long-term exposure to elevated temperature results in a marked reduction in stiffness as well as strength and toughness, the effect usually being greater

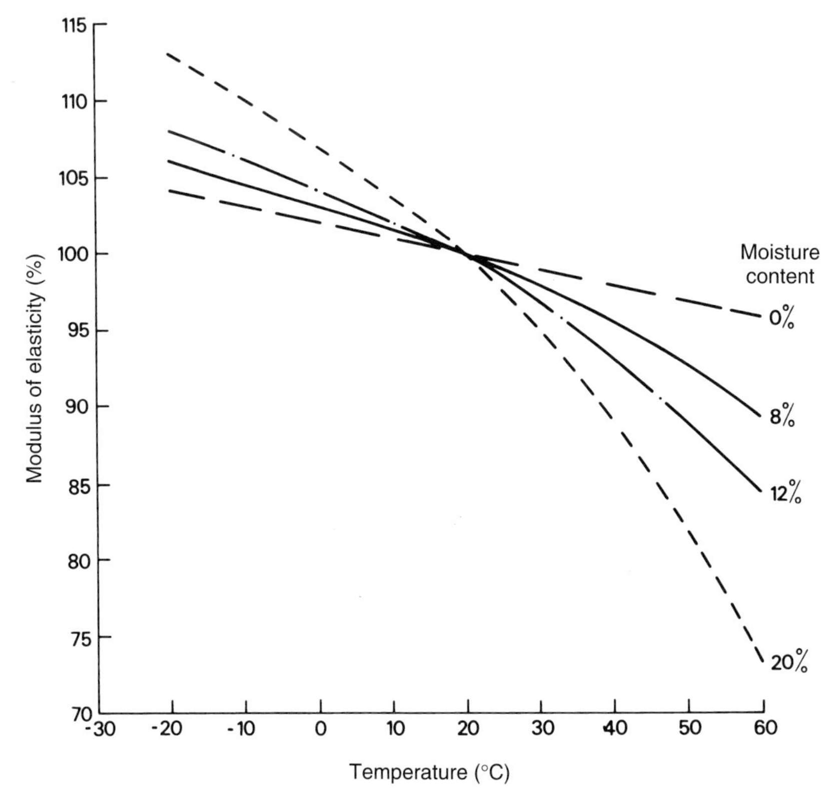

Figure 6.14 The interaction of temperature and moisture content on the modulus of elasticity. Results are averaged for six species of timber and the modulus at 20 °C and 0% moisture content is taken as unity. (© BRE.)

in hardwoods than in softwoods. Even exposure to cyclic changes in temperature over long periods will result in a loss of stiffness (Moore, 1984).

6.2.6 *Bulk modulus*

This is a measure of the cubic compressibility of timber under a uniform hydrostatic pressure and can be obtained from the elastic compliances by the equation:

$$K = \frac{1}{P}\frac{dV}{V} = s_{11} + s_{22} + s_{33} + 2(s_{23} + s_{31} + s_{12}) \tag{6.24}$$

where K is the isothermic cubic compressibility, P is the uniform hydrostatic pressure, and dV/V is the relative change of volume.

From equation (6.24) it follows that

$$\nu_K = \frac{s_{31} + s_{12} + s_{23}}{s_{11} + s_{22} + s_{33}} \tag{6.25}$$

where ν_K is the bulk Poisson's ratio, which for timber has been found to lie between 0.17 and 0.31.

6.3 Viscoelastic deformation

In the introduction to this chapter, timber was described as being neither truly elastic in its behaviour nor truly viscous, but rather a combination of both states; such behaviour is usually described as viscoelastic and, in addition to timber, materials such as concrete, bitumen and the thermoplastics are also viscoelastic in their response to stress.

Viscoelasticity infers that the behaviour of the material is time dependent; at any instant in time under load its performance will be a function of its past history. Now if the time factor under load is reduced to zero, a state which we can picture in concept but never attain in practice, the material will behave truly elastically, and we have seen in Section 6.2 how timber can be treated as an elastic material and how the principles of orthotropic elasticity can be applied. However, where stresses are applied for a period of time, viscoelastic behaviour will be experienced and, although it is possible to apply elasticity theory with a factor covering the increase in deformation with time, this procedure is at best only a first approximation.

In a material such as timber, time-dependent behaviour manifests itself in a number of ways of which the more common are *creep*, *relaxation*, *damping capacity*, and the dependence of strength on *duration of load*. When the load on a sample of timber is held constant for a period of time, the increase in deformation over the initial instantaneous elastic deformation is called creep and Figure 6.1 illustrates not only the increase in creep with time, but also the subdivision of creep into a reversible and an irreversible component, of which more will be said in a later section.

Most timber structures carry a considerable dead load and the component members of these will undergo creep; the dip towards the centre of the ridge of the roof of very old buildings bears testament to the fact that timber does creep. However, compared with thermoplastics and bitumen, the amount of creep in timber is appreciably lower. Although creep behaviour in timber has been recognised for many decades, considerably less research has been carried out on this property compared with the other mechanical properties: this is due to a large extent to the treatment of timber in structural analyses as an elastic material with correction factors for the time variable.

Viscoelastic behaviour is also apparent in the form of relaxation where the load necessary to maintain a constant deformation decreases with time; in timber utilisation this has limited practical significance and the area has attracted very little research. Damping capacity is a measure of the fractional energy converted to heat compared with that stored per cycle under the influence of mechanical vibrations; this ratio is time dependent. A further manifestation of viscoelastic behaviour is the apparent loss in strength of timber with increasing duration of load (this feature is discussed in detail in Section 7.6.12.2 and illustrated in Figure 7.12).

6.3.1 Creep

6.3.1.1 Test methods

Although a simple test in concept, the creep test necessitates the solution of many problems in relation to both loading and recording if high degrees of accuracy and reproducibility are to be achieved. Not only must the load be applied in as near instantaneous time as possible, but it must be held constant for the duration of the test. If dead weights are used their handling is a considerable task, though the quantity can be reduced by applying the load through a lever arm. The deformation of timber under load is affected by temperature and is particularly sensitive to changes in relative humidity. Because of the small deflections obtained, their measurement must be determined with a high degree of accuracy, at least of the order of 0.01 mm.

Deformation can be recorded manually by employing dial gauges of the necessary accuracy, or automatically by using either strain gauges or potentiometers. Use of the former is restricted due to the stiffening effect which the adhesive induces on the surface of the timber and to uncertainties over their time stability. Linear potentiometers, though not without operational problems, are usually employed for test work on timber.

The duration of test can vary from a few hours to several years; greater accuracy in predicting future behaviour is obtained from tests of longer duration.

Although creep in timber under tensile and compressive loading has been measured occasionally, much more attention has been given to creep in bending, due to the high proportion of structural timber that is loaded in this mode.

It is possible to quantify creep by a number of time-dependent parameters of which the two most common are *creep compliance* (known also as *specific creep*) and *relative creep* (known also as the *creep coefficient*); both parameters are a function of temperature.

Creep compliance (c_c) is the ratio of increasing strain with time to the applied constant stress, i.e.

$$c_c(t, T) = \frac{\text{strain varying}}{\text{applied constant stress}} \tag{6.26}$$

whereas relative creep (c_r) is defined as either the deflection, or more usually, the increase in deflection, expressed in terms of the initial elastic deflection, i.e.

$$c_r(t, T) = \frac{\varepsilon_t}{\varepsilon_0} \quad \text{or} \quad \frac{\varepsilon_t - \varepsilon_0}{\varepsilon_0} \tag{6.27}$$

where ε_t is the deflection at time t, and ε_0 is the initial deflection.

Relative creep has also been defined as the change in compliance during the test expressed in terms of the original compliance.

A method for the determination of creep in wood-based panels is given in ENV 1156 (1999).

6.3.1.2 Creep relationships

In both timber and timber products such as plywood or chipboard (particleboard), the rate of deflection or creep slows down progressively with time (Figure 6.15); the creep is frequently plotted against log time and the graph assumes an exponential shape. Results of creep tests can also be plotted as relative creep against log time or as creep compliance against stress as a percentage of the ultimate short-time stress.

In Section 6.2.4 it was shown that the degree of elasticity varied considerably between the horizontal and longitudinal planes. Creep, as one particular manifestation of viscoelastic behaviour, is also directionally dependent. In tensile stressing of longitudinal sections produced with the grain running at different angles, it was found that relative creep was greater in the direction perpendicular to the grain than it was parallel to the grain. Furthermore, in those sections with the grain running parallel to the longitudinal axis, the creep compliance and relative creep across the grain was not only less than the parallel value but also negative (Figure 6.16) . Such anisotropic variation must result in at least some of the Poisson's ratios being time dependent, and actual measurement of some of these ratios has confirmed this view (Schniewind and Barrett, 1972). Such evidence has important consequences; it means that in the stressing of

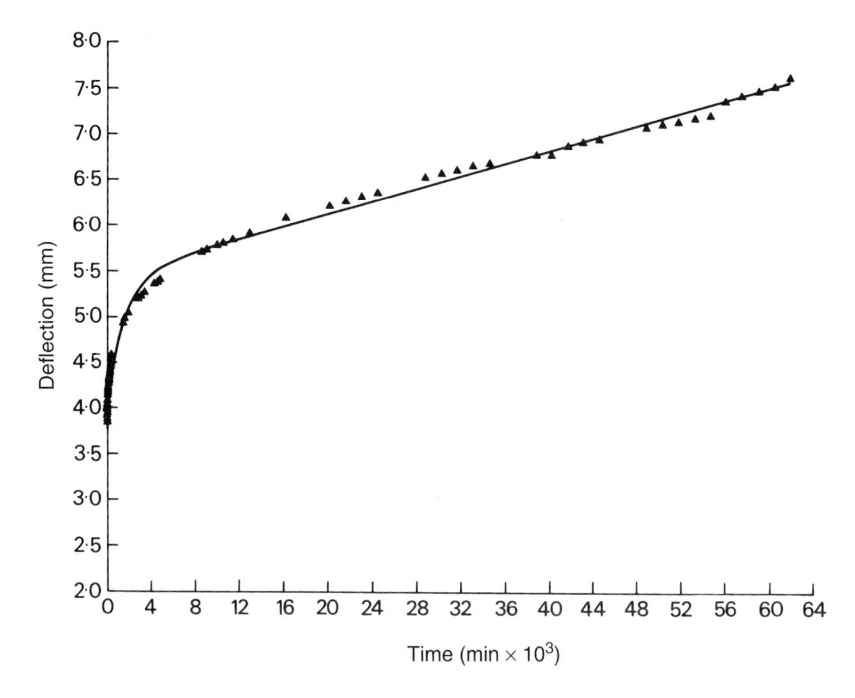

Figure 6.15 The increase in deformation with time of urea–formaldehyde (UF)-bonded chipboard (particleboard) in which the regression line has been fitted to the experimental values using Equation (6.36). (© BRE.)

timber, at least in tension parallel to the grain, an increase in volume occurs and therefore the deformation of timber with time appears to be quite different from those materials which deform at constant volume. The theory of plasticity is frequently applied to these materials, but the concept has rarely been applied to timber.

Timber and wood-based panels are viscoelastic materials, the time dependent properties of which are directionally dependent. The next important criterion is whether they are linear viscoelastic in behaviour. For viscoelastic behaviour to be defined as linear, the instantaneous, recoverable and non-recoverable components of the deformation must vary directly with the applied stress. An alternative definition is that the creep compliance or relative creep must be independent of stress and not a function of it.

Timber and wood-based panels exhibit linear viscoelastic behaviour at the lower levels of stressing, but at the higher stress levels this behaviour reverts to being non-linear. Examples of this transition in behaviour are illustrated in Figure 6.17 where for both redwood timber and UF-bonded particleboard the change from linear to non-linear behaviour occurs between the 45% and 60% stress level.

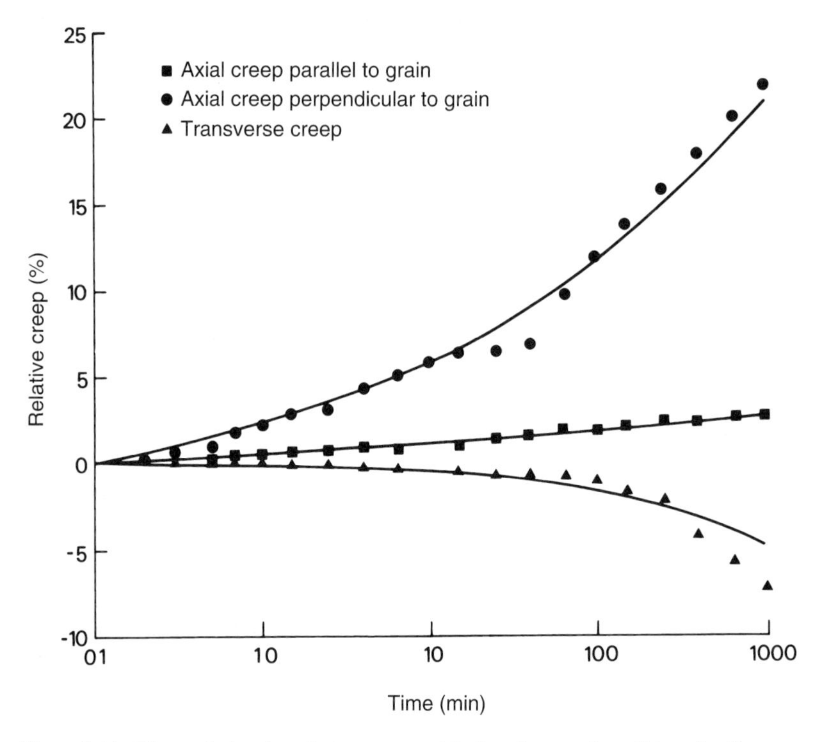

Figure 6.16 The variation in relative creep with time in samples of Douglas fir stressed at different angles to the grain. (From A. P. Schniewind and J. D. Barrett (1972) *Wood Sci. Technol.* **6**, 43–57, reproduced by permission of Springer-Verlag.)

The linear limit for the relationship between creep and applied stress varies with mode of testing, with species of timber or type of panel, and with both temperature and moisture content. In tension parallel to the grain at constant temperature and moisture content, timber has been found to behave as a linear viscoelastic material up to about 75% of the ultimate tensile strength, though some workers have found considerable variability and have indicated a range from 36% to 84%. In compression parallel to the grain, the onset of non-linearity appears to occur at about 70%, though the level of actual stress will be much lower than in the case of tensile strength because the ultimate compression strength is only one third that of the tensile strength. In bending, non-linearity seems to develop very much earlier at about 50–60% (Figure 6.17(a) and (b)); the actual stress levels will be very similar to those for compression.

In both compression and bending, the divergence from linearity is usually greater than in the case of tensile stressing; much of the increased deformation occurs in the non-recoverable component of creep and is associated with progressive structural changes including the development of incipient failure (see Section 7.7.1.2).

Increases not only in stress level, but also in temperature to a limited extent, and moisture content to a considerable degree, result in an earlier onset of non-linearity and a more marked departure from it. For most practical situations, however, working stresses are only a small percentage of the ultimate, rarely approaching even 50%, and it can be safely assumed that timber, like concrete, will behave as a linear viscoelastic material under *normal service conditions*.

6.3.1.3 Principle of superposition

As timber behaves as a linear viscoelastic material under conditions of normal temperature and humidity and at low to moderate levels of stressing, it is possible to apply the *Boltzmann's principle of superposition* to predict the response of timber to complex or extended loading sequences. This principle states that the creep occurring under a sequence of stress increments is taken as the superposed sum of the responses to the individual increments. This can be expressed mathematically in a number of forms, one of which for linear materials is

$$\varepsilon_c(t) = \sum_{1}^{n} \Delta\sigma_i \, c_{ci} \tag{6.28}$$

where n is the number of load increments, $\Delta\sigma_i$ is the stress increment, c_{ci} is the creep compliance for the individual stress increments applied for differing times, $t - \tau_1$, $t - \tau_2$, . . ., $t - \tau_n$ and $\varepsilon_c(t)$ is the total creep at time t. The integrated form is

$$\varepsilon_c(t) = \int_{\tau_1}^{t} c_c(t - \tau) \frac{d\sigma}{d\tau}(\tau) \, d\tau \tag{6.29}$$

In experiments on timber, it has been found that in the comparison of deflections in beams loaded either continuously or repeatedly for 2 or 7 days in every 14, for 6 months at four levels of stress, the applicability of the Boltzmann's principle of superposition was confirmed for stress levels up to 50% (Nakai and Grossman, 1983).

The superposition principle has been found to be applicable even at high stresses in both shear and tension of dry samples. However, at high moisture contents, the limits of linear behaviour in shear and tension appear to be considerably lower, thereby confirming views expressed earlier on the non-linear behaviour of timber subjected to high levels of stressing and/or high moisture content (Section 6.3.1.2).

6.3.1.4 Viscoelasticity: constitutive equations for creep

Provided that some limitations on the conditions for linear behaviour are observed, it is possible to apply linear viscoelastic theory to the long-term deformation of timber. The constitutive equation for a linear anisotropic viscoelastic

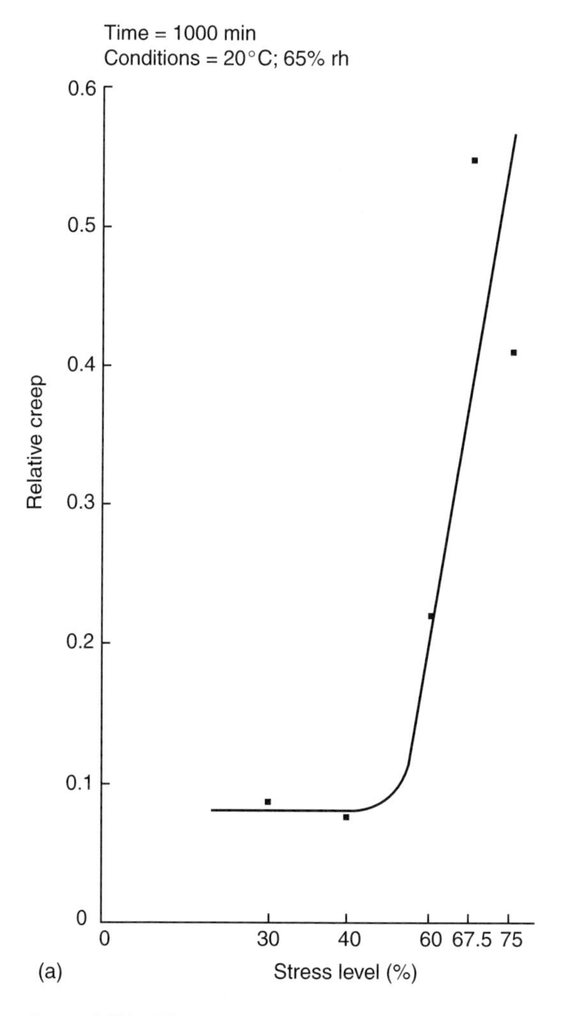

Figure 6.17a The relationship of relative creep to stress level at fixed time periods illustrating the transition from linear to non-linear viscoelastic behaviour in redwood timber (*Pinus sylvestris*). (© BRE.)

material can be written in terms of the creep compliance; as in the treatment of timber as an elastic solid, it is again necessary to treat the circular symmetry of timber as orthotropic. This reduces the 81 compliances of the general case to 36; thermodynamic arguments provide a further reduction to 21, while coincidence between the reference axes and the axes of symmetry leads to a final nine independent compliance parameters. The constitutive equation written in matrix form is therefore

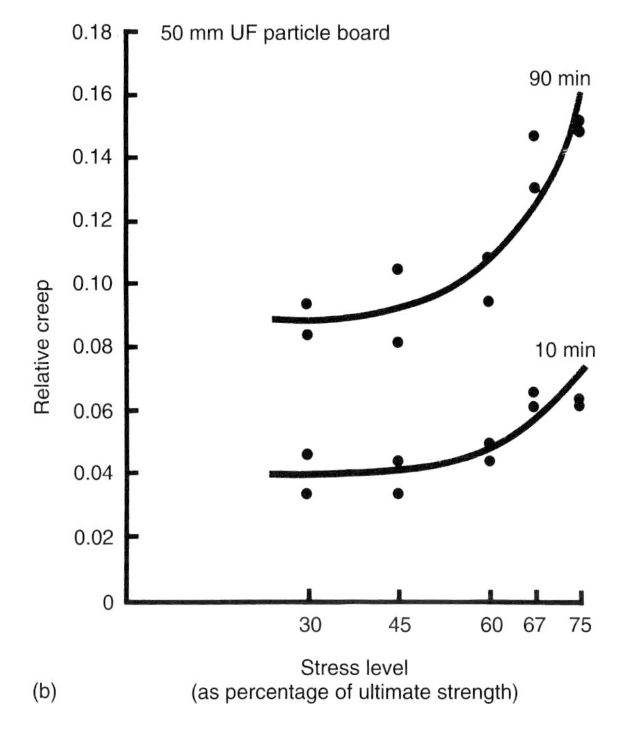

Figure 6.17b The relationship of relative creep to stress level at fixed time periods illustrating the transition from linear to non-linear viscoelastic behaviour in 50 mm UF particleboard. (© BRE.)

$$\begin{bmatrix} \varepsilon_1(t) \\ \varepsilon_2(t) \\ \varepsilon_3(t) \\ \varepsilon_4(t) \\ \varepsilon_5(t) \\ \varepsilon_6(t) \end{bmatrix} = \begin{bmatrix} S_{11}(t) & S_{12}(t) & S_{13}(t) & 0 & 0 & 0 \\ S_{21}(t) & S_{22}(t) & S_{23}(t) & 0 & 0 & 0 \\ S_{31}(t) & S_{32}(t) & S_{33}(t) & 0 & 0 & 0 \\ 0 & 0 & 0 & S_{44}(t) & 0 & 0 \\ 0 & 0 & 0 & 0 & S_{55}(t) & 0 \\ 0 & 0 & 0 & 0 & 0 & S_{66}(t) \end{bmatrix} \begin{bmatrix} \sigma_1 \\ \sigma_2 \\ \sigma_3 \\ \sigma_4 \\ \sigma_5 \\ \sigma_6 \end{bmatrix} \quad (6.30)$$

and, as such, is analogous to equation (6.11) for the elastic behaviour of orthotropic materials.

The nine independent components could be determined experimentally, but to date, generally only isolated compliance components have been determined using uniaxial tests. Schniewind and Barrett (1972), however, measured the principal components of the creep compliance tensor in two planes, thereby permitting the solution of problems of generalised plane stress. In the case of plane stress systems, where it is usual to ignore the very small strain component normal to the plane of applied stress, the constitutive equations, for stress applied in the 1–2 plane, can be written in matrix form as

$$\begin{bmatrix} \varepsilon_1(t) \\ \varepsilon_2(t) \\ \varepsilon_6(t) \end{bmatrix} = \begin{bmatrix} S_{11}(t) & S_{12}(t) & 0 \\ S_{21}(t) & S_{22}(t) & 0 \\ 0 & 0 & S_{66}(t) \end{bmatrix} \begin{bmatrix} \sigma_1 \\ \sigma_2 \\ \sigma_6 \end{bmatrix} \tag{6.31}$$

Equation (6.31) for viscoelastic deformation is therefore analogous to equation (6.22) for elastic deformation. The similarity is continued, for just as it is possible to use transformation equations to obtain the value of strain at some angle to the grain when the timber is treated as deforming elastically, so it is possible to obtain data on creep at some angle to the grain using transformation equations written in terms of the creep compliances.

6.3.1.5 Mathematical modelling of steady-state creep

The relationship between creep and time has been expressed mathematically using a wide range of equations. It should be appreciated that such expressions are purely empirical, none of them possessing any sound theoretical basis. Their relative merits depend on how easily their constants can be determined and how well they fit the experimental results.

One of the most successful mathematical descriptions for creep in timber under constant relative humidity and temperature appears to be the power law, of general form

$$\varepsilon(t) = e_0 + at^m \tag{6.32}$$

where $\varepsilon(t)$ is the time-dependent strain, e_0 is the initial deformation, a and m are material-specific parameters to be determined experimentally ($m = 0.33$ for timber), and t is the elapsed time (Schniewind, 1968; Gressel, 1984).

The prime advantages of using a power function to describe creep is its representation as a straight line on a log–log plot, thereby making onward prediction on a time basis that is much easier than using other models.

Alternative mathematical expressions used to model creep include:

- Exponential functions of the form

  $$\varepsilon_c = a(1 - \exp(-bt)) \tag{6.33}$$

  where ε_c is the *additional* time-dependent strain, and a and b again signify material parameters which require to be determined experimentally.
- Semilogarithmic functions of the form

  $$\varepsilon_c = a + b \log t \tag{6.34}$$

where a and b are as above.

- Second-degree polynominals of the form

$$\varepsilon_c = a + b \log t + c (\log t)^2 \tag{6.35}$$

where a, b and c are again material parameters.

In very general terms, the efficacy of the modelling process decreases from equation (6.32) to (6.35).

Alternatively, creep behaviour in timber, like that of many other high polymers, can be interpreted with the aid of mechanical (rheological) models comprising different combinations of springs and dashpots (a piston in a cylinder containing a viscous fluid); the springs act as a mechanical analogue of the elastic component of deformation, while the dashpots simulate the viscous or flow component. When more than a single member of each type is used, these components can be combined in a wide variety of ways, though only one or two will be able to describe adequately the creep and relaxation behaviour of the material.

The simplest linear model that successfully describes the time-dependent behaviour of timber under constant humidity and temperature for short periods of time is the four-element model illustrated in Figure 6.18; the central part of the model will be recognised as a Kelvin element. To this unit has been added in series a second spring and dashpot. The strain at any time t under a constant load is given by the equation

$$Y = \frac{\sigma}{E_1} + \frac{\sigma}{E_2}\left[1 - \exp\left(\frac{-tE_2}{\eta_2}\right)\right] + \frac{\sigma t}{\eta_3} \tag{6.36}$$

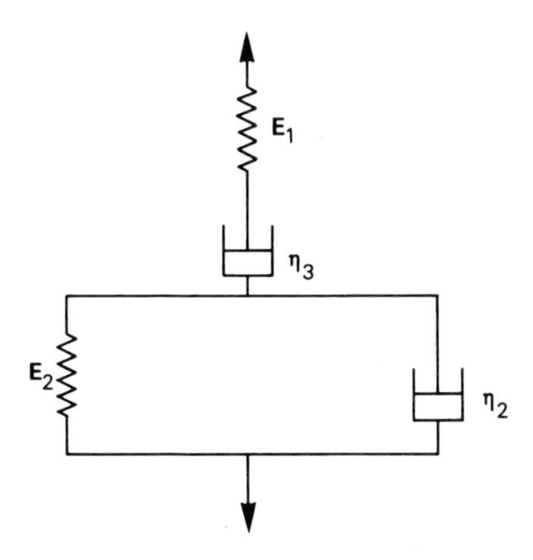

Figure 6.18 Mechanical analogue of the components of creep; the springs simulate elastic deformation and the dashpots viscous flow. The model corresponds to equation 6.36. (© BRE.)

where Y is the strain at time t, E_1 is the elasticity of spring 1, E_2 is the elasticity of spring 2, σ is the stress applied, η_2 is the viscosity of dashpot 2, η_3 is the viscosity of dashpot 3.

The first term on the right-hand side of equation (6.36) represents the instantaneous deformation, the second term describes the delayed elasticity and the third term the plastic flow component. Thus, the first term describes the elastic behaviour while the combination of the second and third terms accounts for the viscoelastic or creep behaviour. The response of this particular model will be linear and it will obey the Boltzmann superposition principle.

The degree of fit between the behaviour described by the model and experimentally derived values can be exceedingly good. An example is illustrated in Figure 6.15, where the degree of correlation between the fitted line and experimental results for creep in bending of urea–formaldehyde chipboard (particleboard) beams was as high as 0.941.

A much more demanding test of any model is the prediction of long-term performance from short-term data. For timber and the various board materials, it has been found necessary to make the viscous term non-linear in these models where accurate predictions of creep (± 10%) are required for long periods of time (over 5 years) from short-term data (6–9 months) (Dinwoodie *et al.* 1990a). The deformation of this non-linear mathematical model is given by

$$Y = \beta_1 + \beta_2[1 - \exp(-\beta_3 t)] + \beta_4 t^{\beta_5} \tag{6.37}$$

where β_5 is the viscous modification factor with a value $0 < b < 1$.

An example of the successful application of this model to predict the deflection of a sample of cement-bonded particleboard (CBPB) after 10 years from the first 9 months of data is illustrated in Figure 6.19.

A further analysis of the data used by Dinwoodie *et al.* (1990a), but with data now covering 12 years of loading, indicated that the model, though still an excellent descriptive tool, had limitations as a predictive tool when the variability in the derived model parameters was taken into account.

By developing a means of assessing the variability occurring in the derived model parameters using the technique of bootstrapping, it was shown that acceptable fits were obtained for only three out of the 20 data sets; this resulted primarily from the model fit predicting an unrealistically large viscous contribution even during the early stages of the experiment (Mundy *et al.*, 1998). The level of prediction was improved considerably by using an alternative method for the derivation of the model parameters, based on the theory behind the mechanical analogue. Thus, this new method constrains the coefficients of the model to give a more realistic ratio between the viscoelastic and viscous components.

Rheological models have proved to be a useful vehicle for the application of *chemical kinetics* to the understanding of creep behaviour by explaining how internal bond strengths change as bonds break and move to more stable

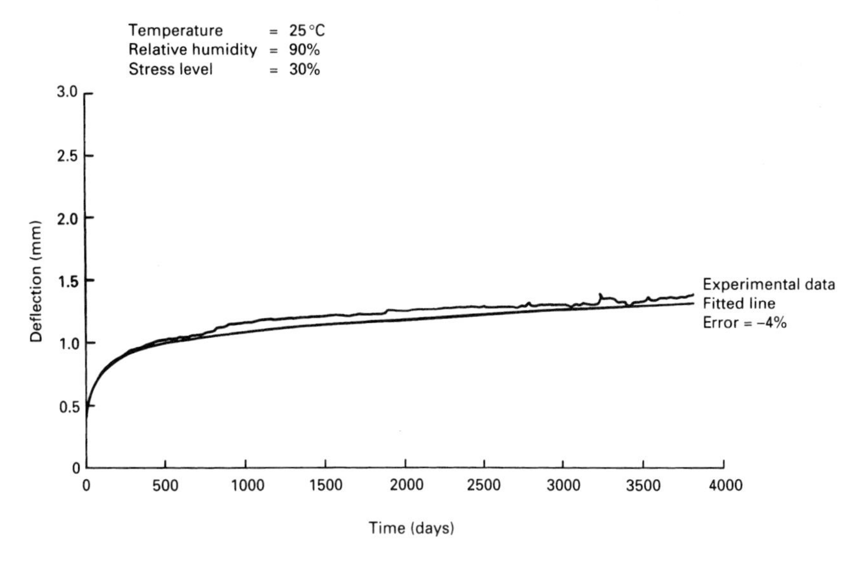

Figure 6.19 Fit of equation (6.37) to the first 36 weeks of creep data for one test piece of cement-bonded particleboard, loaded in three-point bending for 10 years. Error in the predicted deflection at 10 years compared with actual = −4% (© BRE.)

configurations (Caulfield, 1985). Van der Put (1989) applied a three-element non-linear rheological model based on a single energy barrier to describe creep and stress relaxation tests on wood using different test configurations; the model fitted the data well, but the experiment was limited to only short-term data. More recently, Bonfield *et al.* (1996) described two models developed from chemical kinetic theory and validated by experimentation that describe time dependent deformation in wood; one of these models is based on a rheological approach. This work indicated that each deformation element requires energy equivalent to the dissociation of from four to six hydrogen bonds in order to move a distance of between one and two cellobiose units.

For both wood and wood-based panel products, mention was made earlier of the need to use non-linear models in order to accurately describe and predict long-term deformation under low levels of stressing, or short-term deformation under high levels of stressing. It would appear that linear models should only be applied under the application of low levels of stress (<<50%) under steady-state temperature and humidity for short periods of time (see Section 6.3.1.2).

Various non-linear viscoelastic models have been developed and tested over the years ranging from the fairly simple early approach by Ylinen (1965) in which a spring and a dashpot in his rheological model are replaced by non-linear elements, to the much more sophisticated model by Tong and Ödeen (1989a) in which the linear viscoelastic equation is modified by the introduction of a non-linear function either in the form of a simple power function, or by using the sum of an exponential series corresponding to 10 Kelvin elements in series with a single spring. (The development of models for unsteady-state moisture content are described in Section 6.3.1.8).

6.3.1.6 Reversible and irreversible components of creep

In timber and many of the high polymers, creep under load can be subdivided into reversible and irreversible components; passing reference to this was made in Section 6.1 and the generalised relationship with time was depicted in Figure 6.1. The relative proportions of these two components of total creep appears to be related to stress level and to prevailing conditions of temperature and moisture content.

The influence of level of stress is clearly illustrated in Figure 6.20, where the total compliance at 70% and 80% of the ultimate stress for hoop pine in compression is subdivided into the separate components. At the 70% stress level the irreversible creep compliance accounts for about 45% of the total creep compliance, whereas at 80% of the ultimate the irreversible creep compliance has increased appreciably to 70% of the total creep compliance at the longer periods of time, though not at the shorter durations. Increased moisture content and increased temperature will also result in an enlargement of the irreversible component of total creep.

Reversible creep is frequently referred to in the literature as delayed elastic or primary creep and in the early days was ascribed to either polymeric uncoiling or the existence of a creeping matrix. Owing to the close longitudinal association of the molecules of the various components in the amorphous regions, it appears unlikely that uncoiling of the polymers under stress can account for much of the reversible component of creep.

The second explanation of reversible creep utilises the concept of time-dependent two-stage molecular motions of the cellulose, hemicellulose and the lignin constituents. The pattern of molecular motion for each component is dependent on that of the other constituents. It has been shown that the difference in directional movement of the lignin and non-lignin molecules results in considerable molecular interference such that stresses set up in loading can be transferred from one component (a creeping matrix) to another component (an attached, but non-creeping structure). It is postulated that the lignin network could act as an energy sink, maintaining and controlling the energy set up by stressing (Chow, 1973).

Irreversible creep, also referred to as viscous, plastic or secondary creep, has been related to either time-dependent changes in the active number of hydrogen bonds, or to the loosening and subsequent remaking of hydrogen bonds as moisture diffuses through timber with the passage of time (Gibson,1965). Such diffusion can result directly from stressing. Thus Barkas (1945) found that when timber was stressed in tension it gained in moisture content, and conversely

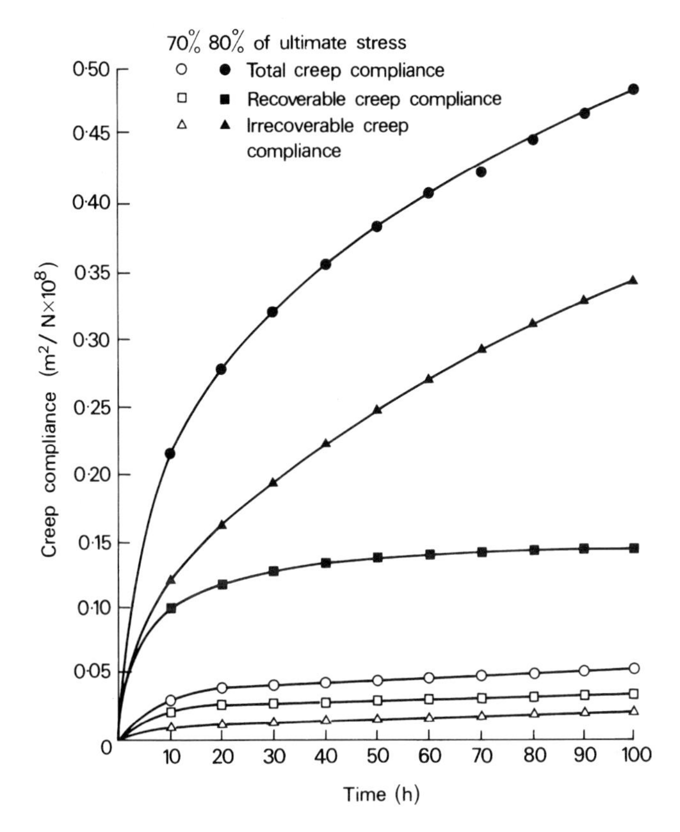

Figure 6.20 The relative proportions of the recoverable and irrecoverable creep compliance in samples of hoop pine (*Araucaria cunninghamii*) stressed in bending. (From R.S.T. Kingston and B. Budgen (1972) *Wood Sci. Technol.*, **6**, 230–238, reproduced by permission of Springer-Verlag.)

when stressed in compression its moisture content was lowered. It is argued, though certainly not proved, that the movement of moisture by diffusion occurs in a series of steps from one adsorption site to the next, necessitating the rupture and subsequent reformation of hydrogen bonds. The process is viewed as resulting in loss of stiffness and/or strength, possibly through slippage at the molecular level. Recently, however, it has been demonstrated that moisture movement, while affecting creep, can account for only part of the total creep obtained, and this explanation of creep at the molecular level warrants more investigation. Certainly, not all the observed phenomena support the hypothesis that creep is due to the breaking and remaking of hydrogen bonds under a stress bias. At moderate to high levels of stressing, particularly in bending and compression parallel to the grain, the amount of irreversible creep is closely associated with the development of kinks in the cell wall (Hoffmeyer and Davidson, 1989). This point is discussed further in Section 7.7.1.2.

Boyd (1982) in a lengthy paper demonstrates how creep under both constant and variable relative humidity can be explained quite simply in terms of stress-induced physical interactions between the crystalline and non-crystalline components of the cell wall. Justification of his viewpoint relies heavily on the concept that the basic structural units develop a lenticular trellis format, containing a water-sensitive gel that changes shape during moisture changes and load applications, thereby explaining creep strains.

Attempts have been made to describe creep in terms of the fine structure of timber and it has been demonstrated that creep in the short term is highly correlated with the angle of the microfibrils in the S_2 layer of the cell wall, and inversely with the degree of crystallinity. However, such correlations do not necessarily prove any causal relationship and it is possible to explain these correlations in terms of the presence or absence of moisture which would be closely associated with these particular variables.

6.3.1.7 *Environmental effects on rate of creep*

TEMPERATURE: STEADY STATE

In common with many other materials, especially the high polymers, the effect of increasing temperature on timber under stress is to increase both the rate and the total amount of creep (Huet *et al.*,1981). Figure 6.21 illustrates a 2.5-fold increase in the amount of creep as the temperature is raised from 20 °C to 54 °C; there is a marked increase in the irreversible component of creep at the higher temperatures.

Various workers have examined the applicability to wood of the time–temperature superposition principle; results have been inconclusive and variable and it would appear that caution must be exercised in the use of this principle (Morlier and Palka, 1994).

TEMPERATURE: UNSTEADY STATE

Cycling between low and high temperatures will induce in stressed timber and panel products a higher creep response than would occur if the temperature was held constant at the higher level, however, this effect is most likely due to changing moisture content as the temperature is changed, rather than to the effect of temperature itself. A small cyclic response under changing temperature, but at constant moisture content, has been recorded by Dinwoodie *et al.* (1992) for timber and board materials. However, this may be an artefact due to changes in rig dimensions with changing temperature, as there does not appear to be any scientific reason why there should be a response to cyclic changes in temperature.

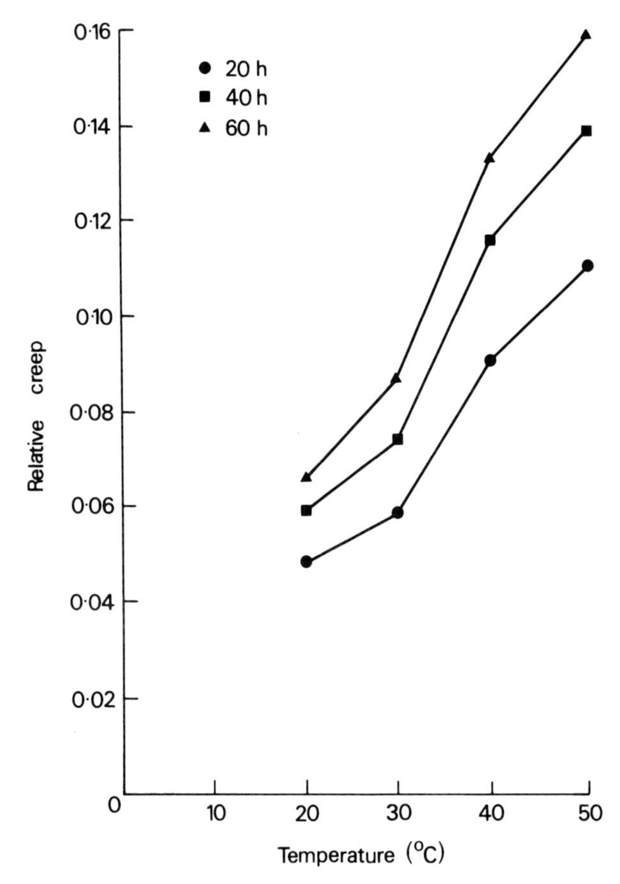

Figure 6.21 The effect of temperature on relative creep of samples of hoop pine (*Araucaria cunninghamii*) loaded in compression for 20, 40 and 60 h. (From R.S.T. Kingston and B. Budgen (1972) *Wood Sci. Technol.*, **6**, 230–238, reproduced by permission of Springer-Verlag.)

MOISTURE CONTENT: STEADY STATE

The rate and amount of creep in timber of high moisture content is appreciably higher than that of dry timber. Clouser (1959) has shown that an increase in moisture content from 6% to 12% increases the deflection at any given stress level by 20%. It is interesting to note the occurrence of a similar increase in creep in nylon when in the wet condition.

Hunt (1999) recorded that the effect of humidity can be treated by the use of an empirical humidity-shift factor curve to be used with an empirical master creep curve. At high moisture contents this logarithmic shift factor increases rapidly; creep at 22% moisture content compared with that at 10% was found to be 32 times as fast.

MOISTURE CONTENT: UNSTEADY STATE

If the moisture content of small timber beams under load is cycled from dry to wet and back to dry again, the deformation will also follow a cyclic pattern. However, the recovery in each cycle is only partial and over a number of cycles the total amount of creep is very large; the greater the moisture differential in each cycle, the higher the amount of creep (Armstrong and Kingston, 1960; Hearmon and Paton, 1964). Figure 6.22 illustrates the deflection that occurs with time in matched test pieces loaded to 3/8 ultimate short-term load where one test piece is maintained in an atmosphere of 93% relative humidity, while the other is cycled between 0 and 93% relative humidity. After 14 complete cycles the latter test piece had broken after increasing its initial deflection by 25 times, whereas the former test piece was still intact having increased in deflection by only twice its initial deflection. Failure of the first test piece occurred, therefore, after only a short period of time and at a stress equivalent to only 3/8 of its ultimate.

It should be appreciated that creep increased during the drying cycle and decreased during the wetting cycle, with the exception of the initial wetting

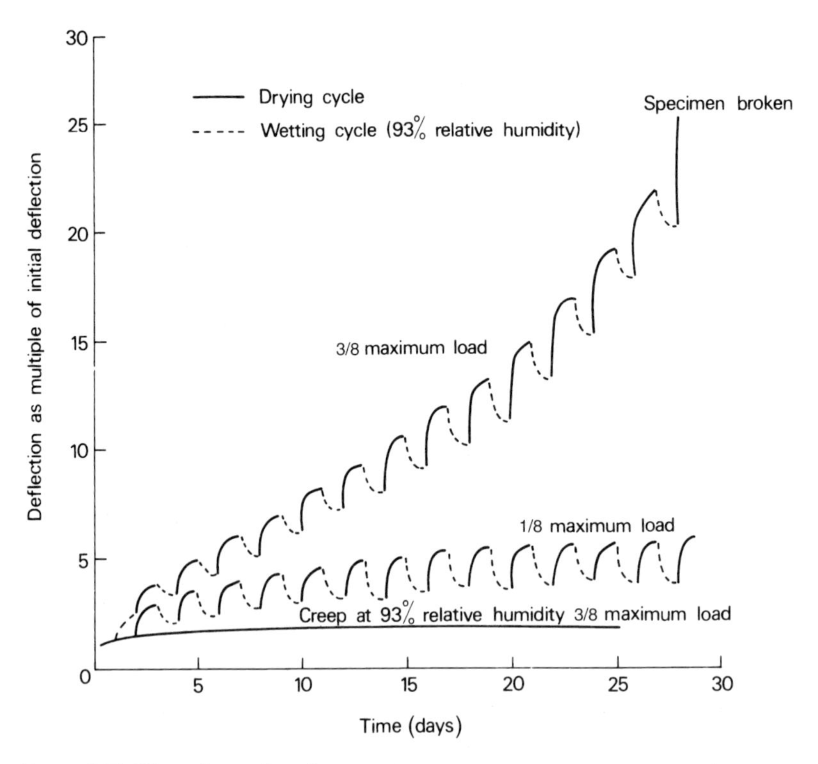

Figure 6.22 The effect of cyclic variations in moisture content on relative creep of samples of beech loaded to 1/8 and 3/8 of ultimate load (© BRE.)

when creep increased. It was not possible to explain the negative deflection observed during adsorption, though the energy for the change is probably provided by the heat of adsorption. The net change at the end of a complete cycle of wetting and drying was considered to be a redistribution of hydrogen bonds, which manifests itself as an increase in deformation of the stressed sample (Gibson, 1965).

More early work was to show that the rate of moisture change affects the rate of creep, but not the amount of creep. This appears to be proportional to the total change in moisture content (Armstrong and Kingston, 1962).

This complex behaviour of creep in timber when loaded under either cyclic or variable changes in relative humidity has been confirmed by a large number of research workers (e.g. Schniewind, 1968; Ranta-Mannus, 1973; Hunt, 1982; Mohager, 1987). However, in board materials the cyclic effect appears to be somewhat reduced (Dinwoodie, 1990b). The amount of creep under variable humidity varies widely among different types with MDF generally having the largest creep deflection (Boehme, 1992; Fernandez-Golfin *et al.*, 1998).

Later test work, covering longitudinal compression and tension stressing as well as bending, indicated that the relationship between creep behaviour and moisture change was more complex than first thought. The results of this work (Hunt, 1982; Figure 6.23) indicated that there are three separate components to this form of creep:

1. An increase in creep follows a decrease in moisture content of the sample, as described previously.
2. An increase in creep follows any increase in moisture content above the previous highest level reached after loading; three examples can be seen in the middle of the graph in Figure 6.23.
3. A decrease in creep follows an increase in moisture content below the previous highest level reached after loading, as described previously.

It follows from 1. and 2. above that there will always be an initial increase in creep during the initial change in moisture content, irrespective of whether adsorption or desorption is taking place.

Further experimentation has established that the amount of creep depends on the size and rate of the moisture change, and is little affected by its duration or by whether such change is brought about by one or more steps (Armstrong and Kingston, 1962). These findings were to cast doubt on the previous interpretation that such behaviour constituted true creep.

Re-inforcement of that doubt occurred with the publication of results of Arima and Grossman (1978). Small beams of *Pinus radiata* 680 × 15 × 15 mm were cut from green timber and stressed to about 25% of their short-term bending strength. While held in their deformed condition, the beams were allowed to dry for 15 days, after which the retaining clamps were removed and the initial recovery measured. The unstressed beams were then subjected to changes in

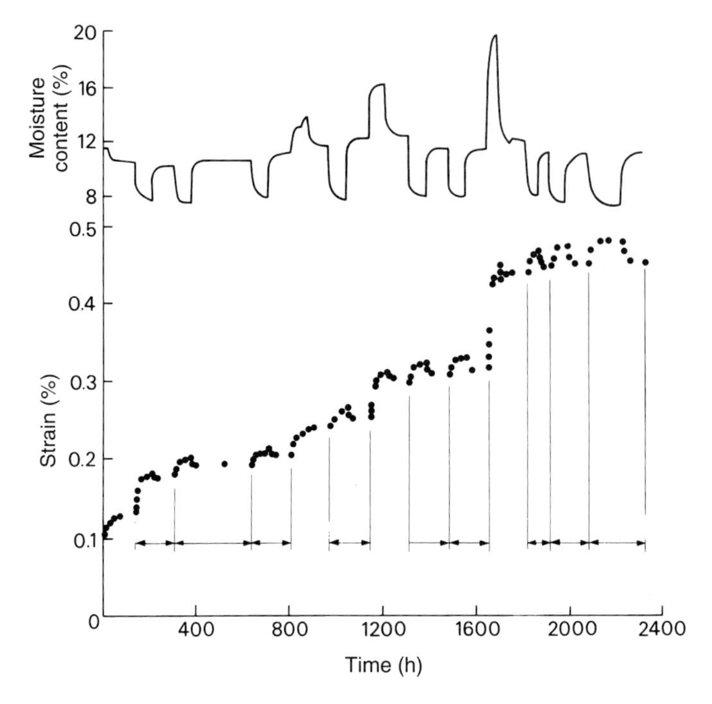

Figure 6.23 Creep deflection under changing moisture content levels for small samples of beech stressed in tension. Note the different responses to increasing moisture content. (From Hunt (1982), *J. Inst. Wood Sci.*, **9** (3), 136–8, reproduced by permission of the Institute of Wood Science.)

relative humidity and Figure 6.24 shows the changes in recovery with changing humidity. Most important is the fact that total recovery was almost achieved; what was thought to have been viscous deformation in the post-drying and clamping stage turned out to be reversible.

These two phenomena – that creep is related to the magnitude of the moisture change and not to time, and that deformation is reversible under moisture change – cast very serious doubts on whether the deformation under changes in moisture content constituted true creep.

It was considered necessary to separate these two very different types of deformation, namely the true viscoelastic creep that occurs under constant moisture content and is directly a function of time, from that deformation which is directly related to the interaction of change in moisture content and mechanical stressing, which is a function of the history of moisture change and is relatively uninfluenced by time. A term of convenience was derived to describe this latter type of deformation, namely *mechano-sorptive behaviour* (Grossman, 1976).

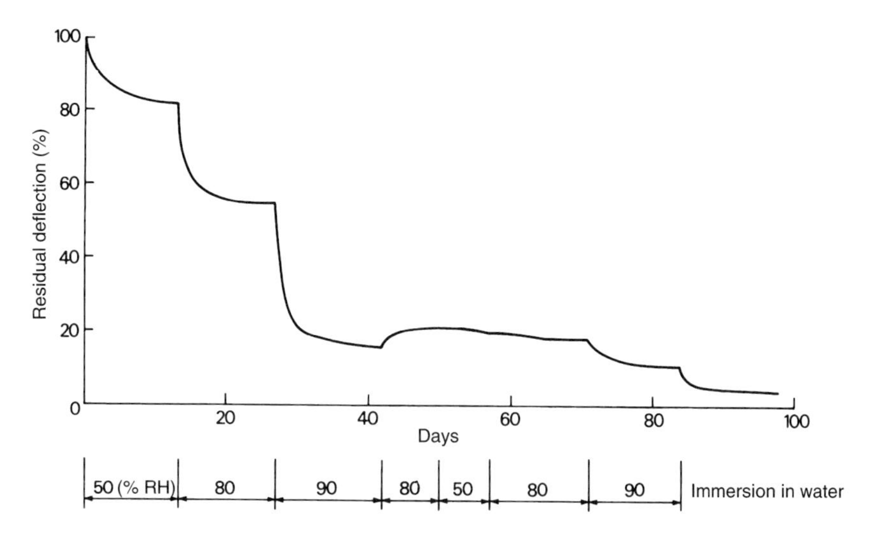

Figure 6.24 The amount of recovery of both viscoelastic and mechano-sorptive deflection that occurred when dried bent beams were subjected to a sequence of humidity changes. (Adapted from T. Arima and P.U.A. Grossman (1978) *J. Inst. Wood Sci.*, **8**, (2), 47–52, by permission of the Institute of Wood Science.)

Changing levels of moisture content, however, will result in changes in the dimensions of timber and an allowance for this must be taken into account in the calculation of mechano-sorptive deformation. Thus,

$$\varepsilon_{m} = \varepsilon_{vc} + \varepsilon_{ms} + \varepsilon_{s} \tag{6.38}$$

hence

$$\varepsilon_{ms} = \varepsilon_{m} - (\varepsilon_{vc} + \varepsilon_{s}) \tag{6.39}$$

where ε_{m} is the total measured strain, ε_{vc} is the normal time (constant moisture content) viscoelastic creep, ε_{ms} is the mechano-sorptive strain under changing moisture content, and ε_{s} is the swelling or shrinkage strain of a matched zero-loaded control test piece. One of the merits of equation (6.39) is that it is a difference equation, which therefore considerably increases the accuracy requirements of any measurements taken to estimate the amount of mechano-sorptive behaviour.

Mechano-sorptive behaviour is linear only at low levels of stress. In the case of compression and bending the upper limit of linear behaviour is of the order of 10%, whereas in tension it is slightly higher (Hunt, 1980).

Susceptibility to mechano-sorptive behaviour is positively correlated with elastic compliance, microfibrillar angle of the cell wall and dimensional change rates (Hunt, 1994). Thus, both juvenile wood and compression wood have been shown to creep much more (up to five times greater) than adult wood.

The swelling or shrinkage strain ε_s, sometimes referred to as pseudo-creep and which is manifest during continued moisture cycling by an increase in deflection during desorption and a decrease during adsorption, has been ascribed to differences in the normal longitudinal swelling and shrinkage of wood; a tensile strain results in a smaller shrinkage coefficient and a compression strain results in a larger one (Hunt and Shelton, 1988).

6.3.1.8 Modelling of deformation under variable moisture content

The requirement for a model of mechano-sorptive behaviour was originally set out by Schniewind (1966) and later developed by Grossman (1976) and Mårtensson (1992). More recently, the list of requirements that must be satisfied for such a model has been reviewed and considerably extended by Hunt (1994); aspects of sorption are now included on the list.

Many attempts have been made since the late 1970s to develop a model for mechano-sorptive behaviour. A few of these models have been explanatory or descriptive in nature, but most of them have been either purely mathematical, with the aim of producing a generalised constitutive equation, or partly mathematical, where the derived equation is linked to some physical phenomenon, or change in structure of the timber under stress. Some examples of these models are provided below.

EXPLANATORY MODELS

These seek relationships at the molecular, ultrastructural or microscopical levels. Reference has been made earlier in this chapter to the concept of bond breakage and subsequent reformation to explain the phenomenon of viscoelastic creep; this concept has also been applied to mechano-sorptive behaviour (Gibson, 1965; Grossman, 1978). The marked reduction in the mechano-sorptive effect in timber which has undergone acetylation or formaldehyde cross-linking (Section 4.2.9) certainly supports the concept that mechano-sorptive behaviour is associated with the breaking and remaking of the hydrogen bonds under stress and change in moisture level. This concept has been further developed by Hunt and Gril (1996) by linking it to the theory of physical ageing, that is the time-dependent approach of a polymer to thermodynamic equilibrium. Such an approach implies that mechano-sorptive creep is not totally independent of time.

It was described in Section 6.3.1.6 on viscoelastic creep how Boyd (1982) had explained creep in terms of stress-induced interactions between crystalline and non-crystalline components of the cell wall by modelling the microfibrils as a trellis that had lenticular openings filled with an amorphous matrix substance. Boyd demonstrated that this model is also capable of explaining mechano-sorptive behaviour; loss in moisture results in shrinkage of the hydrophilic gel inside the microfibrillar trellis, thereby reducing lateral support and causing severe deflection of the microfibrils. These may deflect to such an extent that they make contact with the matrix and make it flow.

Hoffmeyer and Davidson (1989) explain mechano-sorptive behaviour in loading in axial compression at high levels of stress on the formation of *kinks* (slip-planes) in the cell wall. It is suggested that the number of kinks is proportional to the amount of moisture change; longitudinal shrinkage and swelling increase in proportion to the number of kinks. The elastic, viscoelastic and plastic properties of the timber change in proportion to the number of kinks and thus in proportion to the amount of moisture change. A mathematical model is proposed.

PURE MATHEMATICAL MODELS

Many of these models were developed for the three-dimensional case, though they are mainly applied only in the uniaxial case. They are based on the concept that the total strain comprises four components – an elastic strain, a creep strain covering normal time-dependent viscoelastic deformation, a swelling or shrinkage strain, and lastly a strain that covers mechano-sorptive behaviour. The strain terms are then generally treated by associating a separate differential equation or hereditary integral equation to each one. Usually Maxwell, though sometimes Kelvin rheological units, are employed in model formation. Many of the quantitative models assume for simplicity that the terms for the various phenomena can be added.

One of the first attempts to produce a quantitative description of mechano-sorptive behaviour was that by Ranta-Maunus (1973) who proposed a theory of hydroviscoelasticity which describes a functional relationship between deformation and stress, time, temperature and moisture content. It makes the important assumption that some component or components of the timber act as a memory with regard to those four variables. Thus

$$\varepsilon(t) = \sum_{\Psi=0}^{t} [\sigma(\Psi), \mu(\Psi), T(\Psi)] \tag{6.40}$$

where σ is the stress, μ is the moisture content, T is the temperature, and Ψ are the time coordinates. This equation was expanded in a Frechet series containing Kernel functions and was tested against experimental evidence for plywood and wood veneer; reasonable fit between theory and practice was obtained. Later Ranta-Maunus (1989) published an equation based on the differential form which indicated that mechano-sorptive deformation was again linear with moisture change, but had coefficients that took different values depending on the type of moisture change:

$$J = J_{\mathrm{E}} + J_{\mathrm{N}} + J_{\mathrm{EO}} \sum_{i=1}^{n} a(u_i - u_{i-1}) \tag{6.41}$$

where J_E is the elastic compliance, J_N is the 'normal' creep compliance integrated over the moisture contents and times of the test, J_{EO} is the reference elastic compliance at 0% moisture content, a is a hydroviscoelastic constant taking the values a^- or a^+ according to whether the moisture content u_i is less than or greater than u_{i-1}, respectively.

A somewhat similar approach to constitutive modelling to that above was adopted by Mårtensson (1988) to describe mechano-sorptive behaviour of hardboard under a uniaxial tensile stress. Thus

$$\dot{\varepsilon}_x = \frac{\dot{\sigma}_x}{E(w)} + \dot{\varepsilon}_c + \dot{\varepsilon}_s \tag{6.42}$$

where $\dot{\varepsilon}_x$ is the total strain rate, $\dot{\sigma}_x/E(w)$ is the elastic strain rate (where E is dependent on moisture content), $\dot{\varepsilon}_c$ is the creep strain rate, and $\dot{\varepsilon}_s$ is the stress-dependent moisture induced strain rate, given by

$$\dot{\varepsilon}_s = \dot{\varepsilon}_{so} + k\frac{\sigma_x}{\sigma_u}|\dot{\varepsilon}_{so}|$$

where $\dot{\varepsilon}_{so}$ is the free moisture-induced strain rate measured on a non-loaded specimen, σ_u is the ultimate stress in uniaxial tension, and k is a material parameter describing mechano-sorptive behaviour.

One of the problems of such an approach is the determination of k. The model showed relatively good agreement with behaviour under constant stress, but much poorer agreement with behaviour under constant strain (relaxation).

In a later paper, Mårtensson (1992) working on spruce this time, adopted a more generalised non-linear approach to a complex three-dimensional model which was developed on the basis of continuum mechanics theory. However, the model is generally applied in the uniaxial case. Mårtensson embraced in the model allowances for the fact that the shrinkage or swelling response for timber under load differs from that of a non-loaded piece, and that the magnitude of the strain-rate decreases over repeated cycles. The model was successfully used in studies of the response of wood to moisture variations and simultaneous load.

Under conditions of uniformly-sized relative humidity cycles, Hunt (1989) demonstrated a very good empirical fit to the mechano-sorptive creep compliance with a two-term exponential equation to cover 'characteristic' compliances. Thus,

$$J = J_E + J_0 + J_1(1 - e^{-n/N_1}) + J_2(1 - e^{-n/N_2}) \tag{6.43}$$

where J_E is the elastic compliance, J_0 is a constant representing normal viscoelastic creep, J_1 and J_2 are characteristic compliances, n is the number of cycles, and N_1 and N_2 are characteristic cycle numbers.

By making a correction for the change in the shrinkage–swelling rate and having determined the individual parameters for a few humidity cycles, Hunt could then predict a creep limit; the model requires modification for variable changes in the relative humidity.

Toratti (1990) then proposed a linearized model combining some of the components from each of the models described by Hunt (1989) and Ranta-Maunus (1989).

MATHEMATICAL MODELS RELATED TO WOOD STRUCTURE

Bazant (1985) proposed explanations of mechano-sorptive behaviour embracing the thermodynamics of the diffusion process, whereas Gril (1988) developed thermodynamic theory based on the ultrastructure of wood to describe deformation under changing moisture content. He modelled the ultrastructure as a parallel system of two loose chains (which simulate the matrix of hemicellulose, amorphous cellulose and lignin) and an elastic bar which represented the crystalline cellulosic frame. The same hydrogen bonds are deemed to be involved in both creep and sorption. Numerical simulations show good qualitative agreement with measured data.

Van der Put (1989) also proposed a thermodynamic theory to describe creep in constant or changing environments. His theory is based on the assumption of the breaking and remaking of bonds due to sorption under stress, and it is supported by the necessary equations of molecular deformation kinetics. According to this theory, mechano-sorption is not due to the interaction of creep and moisture change.

In contrast, Mukudai and Yata (1987) explained mechano-sorptive behaviour in terms of slippage between the S_1 and S_2 layers of the cell wall at certain moisture content levels. Their model was a complex two-part rheological model including swelling and shrinkage elements and experiments were conducted at high stress levels.

Hanhijärvi (1995) has proposed a new constitutive equation based on the concept that mechano-sorptive behaviour can be explained by the non-linear coupling of the hygroexpansion (swelling or shrinkage) and the viscoelastic creep. Like Gril (1988), he makes the same fundamental assumption that it is the same hydrogen bonds that are activated in the bond breaking and remaking processes in the two phenomena. The mathematical development is accomplished by modifying the flow equation derived in the theory of deformation kinetics (Van der Put, 1989) to account for creep flow, hygroexpansion and their combined effect, namely mechano-sorptive behaviour. The model used is a generalised Maxwell-type model embracing 10 parallel elements. Calculations with the model show good agreement with experimental results.

The concept of mechanosorptive deformation was first established by Armstrong and Kingston. (1960). Since then, the concept of mechano-sorptive deformation, which is primarily a function of the amount of moisture change, has been clearly separated from that of the viscoelastic creep which is primarily a function of time. Indeed, since the late 1970s the concepts have become more polarised and considerable success has been made in recent times in the derivation of effective constitutive equations for mechano-sorptive deformation.

It now appears that the thinking on creep has gone full circle and we are back to the stage where it was once thought that viscoelastic creep and mechano-sorptive behaviour were but two different manifestations of one basic relationship. The published findings by Gril (1988), Hanhijärvi (1995) and Hunt and Gril (1996) indicating that mechano-sorptive deformation may be accounted for by a coupling between creep and hygroexpansion, together with the recent findings by Hunt (1999) that time-dependent creep and mechano-sorptive behaviour are different means of reaching the same creep result, are certainly contrary to views presented since the late 1970s and give cause for a complete re-evaluation of existing concepts.

Hunt's results led him to a new way of characterising wood creep, by plotting data in the form of strain rate against strain. Solution of this differential equation led to the more normal relationship of strain against time. It was then found that normalisation of both the ordinate and abscissa resulted in a single master creep curve for both juvenile and mature wood from a single sample and, more important, approximately also for all test humidities. The effects of humidity changes require the additional measurement of the increased activity associated with the molecular destabilisation, and its relaxation-time constraint, associated with the physical-ageing phenomenon (thermodynamic equilibrium), the application of which suggests that the speed of moisture change might be important in mechano-sorptive deformation, thereby confirming the existence of a size effect which had been suggested previously by van der Put (1989).

More information on the parameters that should be included in these mathematical models as well as the types of models that have been developed over the years, is available in the excellent texts by Tong and Ödeen (1989b), Hunt (1994), Morlier and Palka (1994) and Hanhijärvi (1995).

6.3.2 Relaxation

Another characteristic of viscoelastic materials is that they will relax with time; i.e. when a sample is stressed for a long period of time, the level of stress necessary to maintain a constant deflection will decrease with time. The process is usually quantified in terms of a relaxation modulus:

$$M_r(t, T) = \frac{\text{stress (varying)}}{\text{applied constant strain}} \tag{6.44}$$

Although the relaxation modulus and the creep compliance are related it must be appreciated that it is only when time (t) is zero that they are exactly reciprocals. Studies of relaxation in timber have been carried out and the factors responsible for this form of time-dependent behaviour are identical with those causing creep. Non-linearity appears to develop early at low levels of initial strain, but fortunately the degree of non-linearity is very small over a large range of initial values and timber can still be treated as a linear viscoelastic material.

References

Standards and specifications

BS 373 (1986) *Methods of testing small clear specimens of timber,* BSI, London.

BS 5820 (1979) *Methods of test for determination of certain physical and mechanical properties of timber in structural sizes*, BSI, London.

EN 408 (1995) *Timber structures – Structural timber and glued laminated timber – determination of some physical and mechanical properties.*

ENV 1156 (1999) *Wood-based panel products – determination of duration of load and creep factors*, BSI, London.

Literature

Arima, T. and Grossman, P.U.A. (1978) Recovery of wood after mechano-sorptive deformation. *J. Inst. Wood Sci.*, **8** (2), 47–52.

Armstrong, L.D. and Kingston, R.S.T. (1960) Effect of moisture changes on creep in wood. *Nature*, **185** (4716), 862–863.

Armstrong, L.D. and Kingston, R.S.T. (1962) The effect of moisture content changes on the deformation of wood under stress. *Aust. J. App. Sci.*, **13** (4), 257–276.

Astley, R.J., Stol, K.A. and Harrington, J.J. (1998) Modelling the elastic constraints of softwood. Part II: The cellular microstructure. *Holz als Roh-und Werkstoff*, **56**, 43–50.

Barkas, W.W. (1945) *Swelling stresses in gels*, Special Report No 6 on Forest Products Research, HMSO, London.

Bazant, Z. (1985) Constitutive equation of wood in variable humidity and temperature. *Wood Sci. Technol.*, **19**, 159–177.

Boehme, C. (1992) Kriechverhalten UF-Verleimter MDF. *Holz als Roh und Werkstoff*, **50**, 158–162.

Bonfield, P.W., Mundy, J., Robson, D.J. and Dinwoodie J.M. (1996) The modelling of time-dependent deformation in wood using chemical kinetics. *Wood Sci. Technol.*, **30**, 105–115.

Boyd, J.D. (1982) An anatomical explanation for viscoelastic and mechano-sorptive creep in wood, and effects of loading rate on strength, in *New perspectives in Wood anatomy*, Ed. P. Bass, Martinus Nijhoff/Dr W. Junk, The Hague, 171–222.

Carrington, H. (1922) The elastic constants of spruce as affected by moisture content. *Aeronautical Journal*, **26**, 462–471.

Caulfield, D.F. (1985) A chemical kinetics approach to the duration of load problem in wood. *Wood and Fiber Sci.*, **17** (4), 504–521.

Cave, I.D. (1968) The anisotropic elasticity of the plant cell-wall. *Wood Sci. Technol.*, **2**, 268–278.

Cave, I.D. (1975) Wood substance as a water-reactive fibre-reinforced composite. *J. Microscopy*, **104** (1), 47–52.
Chow, S. (1973) Molecular rheology of coniferous wood tissues. *Transactions of the Society of Rheology*, **17**, 109–128.
Clouser, W.S. (1959) Creep of small wood beams under constant bending load, Report 2150 Forest Products Laboratory, Madison.
Cowdrey, D.R. and Preston, R.D. (1996) Elasticity and microfibrillar angle in the wood of Sitka spruce. *Proc. Roy. Soc. B*, **166**, 245–272.
Dinwoodie J.M. (1975) Timber – a review of the structure-mechanical property relationship. *J Microscopy* **104** (1), 3–32.
Dinwoodie, J.M., Higgins, J.A., Paxton, B.H. and Robson, D.J. (1990a) Creep in chipboard. Part 7: Testing the efficacy of models on 7–10 years data and evaluating optimum period of prediction. *Wood Sci. Technol.*, **24**, 181–189.
Dinwoodie, J.M., Higgins, J.A., Paxton, B.H. and Robson, D.J. (1990b) Creep research on particle board – 15 years work at UK BRE, *Holz als Roh und Werkstoff*, **48**, 5–10.
Dinwoodie J.M., Higgins J.A., Paxton B.H. and Robson, D.J. (1992) Creep in chipboard. Part 11: The effect of cyclic changes in moisture content and temperature on the creep behaviour of a range of boards at different levels of stressing, *Wood Sci. Technol.*, **26**, 429–448.
Fernandez-Golfin Seco, J.I. and Diez Barra, M.R. (1998) Long-term deformation of MDF panels under alternating humidity conditions. *Wood Sci. Technol.*, **32**, 33–41.
Gerhards, C.C. (1982) Effect of moisture content and temperature on the mechanical properties of wood. An analysis of immediate effects. *Wood & Fiber*, **14** (1) ,4–36.
Gibson, E. (1965) Creep of wood: role of water and effect of a changing moisture content. Nature (*Lond*), **206**, 213–215.
Gressel, P. (1984) Zur Vorhersage des langfristigen Formänderungsverhaltens aus Kurz-Kriechversuchen. *Holz als roh-und Werkstoff*, **42**, 293–301.
Gril, J. (1988) *Une modélisatin du compårtement hygro-rhéologique du bois à partir de sa microstructure*. PhD Thesis, University of Paris.
Grossman, P.U.A. (1976) Requirements for a model that exhibits mechano-sorptive behaviour. *Wood Sci. Technol.*, **10**, 163–168.
Grossman, P.U.A. (1978) Mechano-sorptive behaviour, in Jayne, B.A., Johnson, J.A. and Perkins, R.W. (Eds), General constitutive relations for wood and wood-based materials. Report of July 1976 NSF Workshop at Blue Mountain Lake, Minnowbrook, NY Syracuse University 313–325.
Hanhijärvi, A. (1995) *Modelling of creep deformation mechanisms in wood*: publication **231,** Technical Research Centre of Finland.
Harington, J.J., Booker, R. and Astley, R.J. (1998) Modelling the elastic constraints of softwood. Part 1; The cell-wall lamellae. *Holz als Roh-und Werkstoff*, **56**, 37–41.
Hearmon, R.F.S. (1948). Elasticity of wood and plywood, *Special Report No 7 on Forest Products Research, London.*
Hearmon, R.F.S. (1966) Vibration testing of wood, *For. Prod. J.*, **16** (8), 29–40.
Hearmon, R.F.S. and Paton, J.M. (1964) Moisture content changes and creep in wood, *For. Prod. J.*, **14**, 357–359.
Hoffmeyer, P. and Davidson, R. (1989) Mechano-sorptive creep mechanism of wood in compression and bending. *Wood Sci. Technol.*, **23**, 215–227.
Huet, C., Guitard, D. and Morlier, P. (1981) Le bois en structure, son compartement différé, Report No 470, Institut Technique du Batiment et des Travaux Public.

Hunt, D.G. (1980) A preliminary study of tensile creep of beech with concurrent moisture changes, in *Proceedings of the Third International Conference on Mechanical Behaviour of Materials*, Cambridge, 1979, Eds, K.J. Miller and R.F. Smith, Vol. **3**, 299–308.

Hunt, D.G. (1982) Limited mechano-sorptive creep of beech wood. *J. Inst. Wood Sci.*, **9** (3), 136–138.

Hunt, D. G. (1989) Linerarity and non-linearity in mechano-sorptive creep of softwood in compression and bending. *Wood Sci. Technol.*, **23**, 323–333.

Hunt, D.G. (1994) Present knowledge of mechano-sorptive creep of wood, in *Creep in timber structures*, Ed. P. Morlier, Rilem Report 8, E & F N Spon, London, pp. 73–97.

Hunt, D.G. (1999) A unified approach to creep in wood. *Proc. Roy. Soc. London, A*, accepted for publication.

Hunt, D.G. and Gril, J. (1996) Evidence of a physical ageing phenomenon in wood. *J Mat. Sci. Letters*, 15, 80–92.

Hunt, D.G and Shelton, C.F. (1988) Longitudinal moisture-shrinkage coefficents of softwood at the mechano-sorptive creep limit. *Wood Sci. Technol.*, **22**, 199–210.

Kingston, R.S.T. and Budgen, B. (1972) Some aspects of the rheological behaviour of wood, Part IV: Non-linear behaviour at high stresses in bending and compression *Wood Sci. Technol.*, **6**, 230–238.

Mårtensson, A. (1988) Tensile behaviour of hardboard under combined mechanical and moisture loading. *Wood Sci. Technol.*, **22**, 129–142.

Mårtensson, A. (1992) *Mechanical behaviour of wood exposed to humidity variations,* Report TVBK-1066, Lund Institute of Technology, Sweden.

Mohager, S. (1987) *Studier av krypning hos trä (studies of creep in wood).* Report 1987–1 of the Dept. of Building Materials, The Royal Institute of Technology, Stockholm.

Moore, G.L. (1984) The effect of long term temperature cycling on the strength of wood. *J. Inst. Wood Sci,*. **9** (6), 264–267.

Morlier, P and Palka, L.C. (1994) Basic knowledge, in *Creep in timber structures*, Ed. P. Morlier, Rilem Report 8, E & FN Spon, pp. 9–42.

Mukudai, J. and Yata, S. (1987) Further modelling and simulation of viscoelastic behaviour of wood under moisture change. *Wood Sci. Technol.*, **21**, 49–63.

Mundy, J.S., Bonfield P.W., Dinwoodie, J.M. and Paxton, B.H. (1998) Modelling the creep behaviour of chipboard: the rheological approach. *Wood Sci. Technol.*, **32**, 261–272.

Nakai, T. and Grossman, P.U.A. (1983) Deflection of wood under intermittent loading. Part 1; Fortnightly cycles. *Wood Sci. Technol.*, **17**, 55–67.

Ranta-Maunus, A. (1973) *A theory for the creep of wood with application to birch and spruce plywood*, Publication 4. Technical Research Centre of Finland, Building Technology and Community Development.

Ranta-Maunus, A. (1989) Analysis of drying stresses in timber. *Paperi ja Puu*, **71**, 1120–1122.

Schniewind, A.P. (1966) Uber den Einfluss von Feuchtigkeitsanderungen auf das Kriechen von Buchenholz quer zur Faser unter Berucksichtigung von Temperatur und temperaturanderungen. *Holz als Roh-und Werkstoff*, **24**, 87–98.

Schniewind, A.P. (1968) Recent progress in the study of rheology of wood. *Wood Sci. Technol.*, **2**, 189–205.

Schniewind, A.P. and Barrett, J.D. (1972) Wood as a linear orthotropic viscoelastic material. *Wood Sci. Technol.*, **6**, 43–57.

Sulzberger, P.H. (1947) *The effect of temperature on the strength of wood at various moisture contents in static bending*, Progress Report 7, Project TP 10–3, Commonwealth Scientific and Industrial Research Organisation (Australia), Division of Forest Products.

Tong, L. and Ödeen, K. (1989a) *A non-linear viscoelastic equation for the deformation of wood and wood structure*, Report No. 4 Department of Building Materials, Royal Institute of Technology, Stockholm.

Tong, L. and Ödeen, K. (1989b) *Rheological behaviour of wood structures*, Report No. 3, Department of Building Materials, Royal Institute of Technology, Stockholm.

Toratti, T. (1990) A cross-section creep analysis, in *International Union of Forestry Research Organisations Timber Engineering Conference, St John, Canada.*

Van der Put, T.A.C.M. (1989) *Deformation and damage processes in wood.* PhD thesis, Delft University Press.

Ylinen, A. (1965) Prediction of the time-dependent elastic and strength properties of wood by the aid of a general non-linear viscoelastic rheological model, *Holz als Roh und Werkstoff*, **5**, 193–196.

Chapter 7

Strength and failure in timber

7.1 Introduction

Whereas it is easy to appreciate the concept of deformation primarily because it is something that can be observed, it is much more difficult to define in simple terms what is meant by the *strength* of a material. Perhaps one of the simpler definitions of strength is that it is a measure of the resistance to failure, providing of course that we are clear in our minds what is meant by failure.

Let us start therefore by defining failure. In those modes of stressing where a distinct break occurs with the formation of two fracture surfaces failure is synonymous with rupture of the specimen. However, in certain modes of stressing, fracture does not occur and failure must be defined in some arbitrary way, such as the maximum stress that the sample will endure or, in exceptional circumstances such as compression strength perpendicular to the grain, the stress at the limit of proportionality.

Having defined our end point, it is now easier to appreciate our definition of strength as the natural resistance of a material to failure. But how do we quantify this resistance? This may be done by calculating either the stress necessary to produce failure or the amount of energy consumed in producing failure. Under certain modes of testing it is more convenient to use the former method of quantification as the latter tends to be more limited in application. Before describing some of the more frequently applied modes of loading timber, mention must first be made of the use of standard sizes of sample and the adoption of sampling techniques for the characterisation of strength in different timbers.

7.2 Determination of strength

7.2.1 Test piece size and selection

Although in theory it should be possible to determine the strength properties of timber independent of size, in practice this is found not to be the case. A definite though small size effect has been established and in order to compare the strength of a timber sample with recorded data it is advisable to adopt the standard sizes set out in the literature.

The size of the test piece to be used will be determined by the type of information required. When this has been decided, it will determine the test procedure; standardised test procedures should always be adopted and a choice is available between national, European, or international methods of test (Figure 7.1).

7.2.1.1 Use of small clear test pieces

This size of test piece was originally used for the derivation of working stresses for timber, but in the mid-1970s it was superseded by structural-size timber. However, the small clear test piece still remains valid for characterising new timbers and for the strict academic comparison of wood from different trees or different species, because the use of small knot-free straight-grained perfect test pieces represent the maximum quality of wood that can be obtained. As such, these test pieces are not representative of structural-size timber, with all its imperfections, and several arbitrarily defined reduction factors have to be used in order to obtain a measure of the working stresses of the timber, when small clear test pieces are used (see Section 7.8.2.1).

Two standard procedures for testing small clear test pieces have been used internationally. The original was introduced in the USA as early as 1891 using a test sample of 2 × 2 inches in cross-section. The second, European in origin, employs a test specimen 20 × 20 mm in cross-section. Prior to 1949 the former size was adopted in the UK, but after this date this larger sample was superseded by the

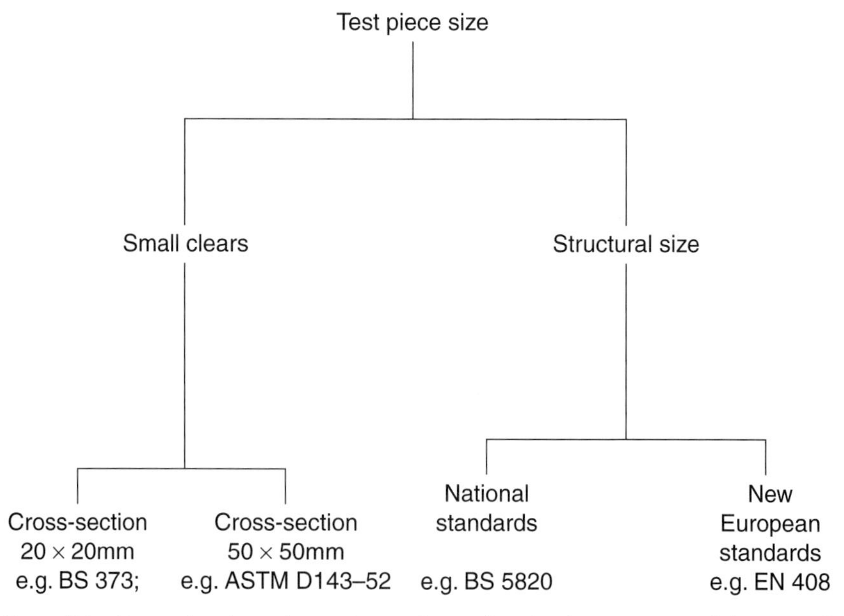

Figure 7.1 Alternative sizes of test piece to be used in the determination of the strength of timber.

smaller, thereby making it possible to obtain an adequate number of test specimens from smaller trees. Because of the differences in size, the results obtained from the two standard procedures are not strictly comparable and a series of conversion values has been determined. With the exception of stiffness, which has a ratio of 1.07 for the mean of the 2 in results divided by the mean of the 20 mm results, most of the strength properties have ratios of from 0.79 to 0.96 (Lavers, 1969).

The early work in the UK on species characterisation employed a sampling procedure in which the test samples were removed from the log in accordance with a cruciform pattern. For 2 × 2 inch samples this method is described fully in ASTM Standard D143-52 (1972) which is still used in the USA and in some of the South American countries. After 1949, the practice in the UK was at first to retain the cruciform sampling scheme, but to modify it slightly. However, this was subsequently abandoned and a method devised applicable to the centre plank removed from a log; 20 × 20 mm sticks, from which the individual test pieces are obtained, are selected at random in such a manner that the probability of obtaining a stick at any distance from the centre of a cross-section of a log is proportional to the area of timber at that distance. However, the cruciform method is retained for use on very small diameter logs. Test samples are cut from each stick eliminating knots, defects and sloping grain; this technique is described fully by Lavers (1969).

7.2.1.2 Use of structural-size test pieces

The use of these larger test pieces reproduces actual service loading conditions and they are of particular value because they allow directly for defects such as knots, splits and distorted grain rather than by applying a series of reduction factors as is necessary with small clear test pieces. However, use of the large pieces is probably more costly.

7.2.2 Standardised test procedures

Europe at the present time is in a transition period in which national test procedures are being replaced by European procedures. Within the first decade of the twenty-first century all national standards relating to the use of timber and panel products in construction will be withdrawn. It is uncertain at this stage whether the testing of small clear test pieces will be retained as a national standard, provided it is not used for the derivation of characteristic values (working stresses). It is interesting to note that many of the European standards (ENs) have now been adopted as international standards (ISOs).

7.2.2.1 Using 20 × 20 mm small clear test pieces

The methods of test in the UK are set out in BS 373 (1986) *Methods of Testing Small Clear Specimens of Timber*. A number of improvements in methods have been introduced over the years though the standard has not been fully revised. The important points in the standard for each test, together with the improvements, are listed below for the more important properties, though it should be appreciated that this is not the complete list of tests.

TENSION PARALLEL TO THE GRAIN

In this test, the test piece is 300 × 20 × 6 mm and waisted in the central region to 6 mm × 3 mm; a test piece under test is illustrated in Figure 6.5 and the calculated stress at rupture is recorded as the ultimate tensile strength. This test is performed only infrequently as the amount of timber loaded in tension under service conditions is quite small. A further reason for the lack of tensile data are the difficulties experienced in performing the tensile test: first, due to the very high tensile strength of timber, it is difficult to grip the material without crushing the grain, especially in low-density timbers; and second, in timber with very high tensile strength, failure is frequently in shear at the end of the waisted region rather than in tension within the waisted region. It is very difficult to conduct the standard tensile test in green timber.

COMPRESSION PARALLEL TO THE GRAIN

The test piece size is 60 × 20 × 20 mm and the load is applied at a rate of 0.01 mm/s through a ball contact and plunger. The maximum strength parallel to the grain is obtained by dividing the maximum load recorded during the test by the cross-sectional area of the test piece.

STATIC BENDING STRENGTH

A test piece 300 × 20 × 20 mm is supported over a span of 280 mm in trunnions carried on roller bearings and the load is applied to the centre of the beam at a constant strain rate of 0.11 mm/s. The orientation of the growth rings is parallel to the direction of loading and an extensometer is usually attached to provide a load–deflection diagram from which is calculated the modulus of elasticity (see Section 6.2.1.3 and equation (6.2)). Three strength properties are usually determined from this test. The first and most important is the *modulus of rupture*, which is a measure of the ultimate bending strength of timber for that size of test piece and that rate of loading. This modulus is actually the equivalent stress in the extreme fibres of the specimen at the point of failure, calculated on the assumption that the simple theory of bending applies. In three-point bending the modulus of rupture (MOR) is given by

$$\text{MOR (N/mm}^2\text{)} = \frac{3PL}{2bd^2} \tag{7.1}$$

where P is the load in newtons, L is the span length in millimetres, b is the width of the beam in millimetres, and d is the thickness of the beam in millimetres.

The second strength parameter is *work to maximum load,* which is a measure of the energy expended in failure and is determined from the area under the load–deflection curve up to the point of maximum load. This parameter is consequently a measure of the toughness of timber, as is also the third parameter, namely *total work*, where the area under the load–deflection curve is taken to complete failure. Both values have units of mm N/mm^3.

IMPACT BENDING

It will be noted above how two energy parameters of the static bending test can provide a measure of the toughness of the material. A number of more direct methods have been employed in timber, many of these borrowed from metallurgy. Thus, both Charpy and Izod methods have been used but, owing to the variability of the structure of timber, the small samples used in these tests can be unrepresentative of the material. Consequently, larger samples have been adopted and tested in a modified Hatt–Turner machine (Figure 7.2). In this test, a test piece $300 \times 20 \times 20$ mm is supported over a span of 240 mm on chair supports radiused to 15 mm. A mass of 1.5 kg, also radiused to 15 mm, is dropped onto the beam (test piece) from increasing heights until failure occurs, or the deflection equals 60 mm.

Criticism has been levelled at this technique principally on the grounds that the test piece is subjected to repeated blows rather than one single blow. Nevertheless the test, though so empirical in nature, appears to give a good indication of the toughness of a timber in practice.

SHEAR PARALLEL TO THE GRAIN

This test is made on a 20 mm cube using a pivoted-arm shear test rig as illustrated in Figure 7.3. The cube is loaded through a ball seating at a rate of 0.0085 mm/s. Tests are made on matched pairs of specimens in the radial and tangential planes and averaged to give an *ultimate shear strength.*

CLEAVAGE

The test piece is $45 \times 20 \times 20$ mm with a transverse groove at one end into which are slotted the tensile grips. The rate of loading is 0.042 mm/s and tests are made on matched pairs of test pieces to give failures in both radial and tangential planes. The failing load is usually expressed as force per unit width of test piece. A similar test in which the test piece is grooved at both ends is used to measure the tensile strength of timber perpendicular to the grain.

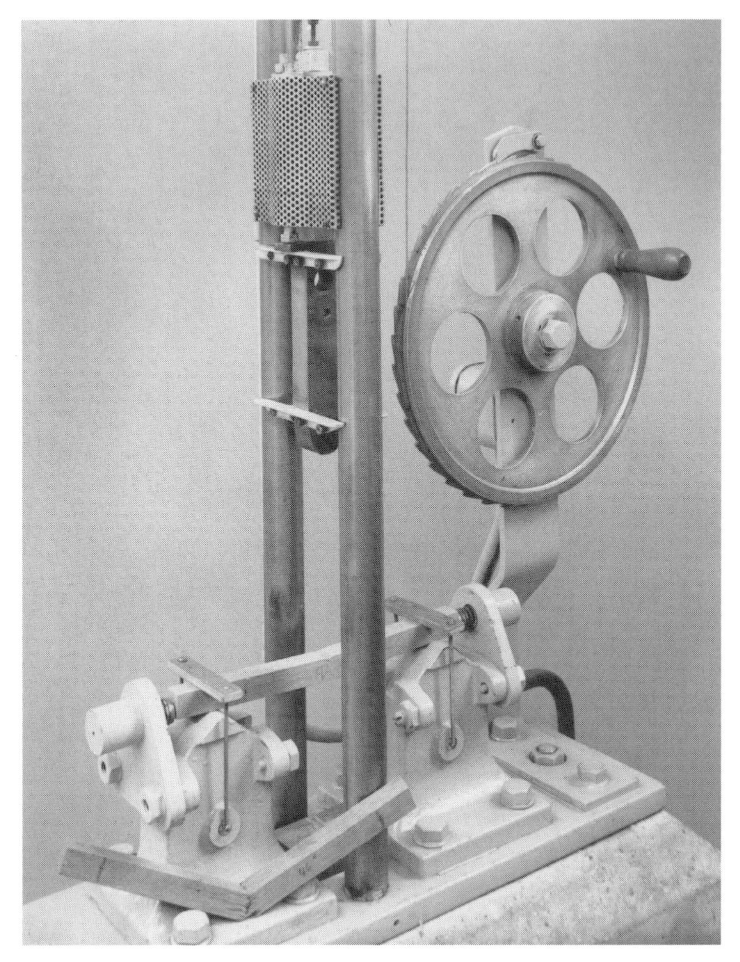

Figure 7.2 Modified Hatt–Turner impact bending test on a 300 × 20 × 20 mm specimen. (© BRE.)

HARDNESS

This is an important property in the use of timber for both domestic and industrial flooring. A specially hardened steel tool rounded to a diameter of 11.3 mm is embedded in the timber to a depth corresponding to half its diameter at a rate of 0.11 mm/s.

Figure 7.3 Shear parallel to the grain using a 20 mm cube. (© BRE.)

7.2.2.2 Using structural-size test pieces to EN 408

BENDING STRENGTH

The test piece normally has a minimum length of 19 times the depth of the cross-section, and is symmetrically loaded at *two* points over a span of 18 times the depth of the cross-section; the distance between the two loading points is six times the depth of the cross-section. The beam is usually tested in four-point bending, in contrast to the three-point loading system used for testing small clear test pieces, thereby achieving a constant bending moment between the two loading points which is more representative of a beam in service.

Load is applied at a constant loading-head movement so adjusted that maximum load is reached within (300 ± 120) s. Bending strength is given by the equation

$$f_{\mathrm{m}} = \frac{aF_{\mathrm{max}}}{2W} \tag{7.2}$$

where f_{m} is the bending strength in newtons per square millimetre, F_{max} is the maximum load achieved in newtons, a is the distance between a loading position and the nearest support in millimetres, and W is the section modulus in millimetres cubed.

TENSION STRENGTH PARALLEL TO THE GRAIN

The test piece is of full cross-section throughout its length, and is of sufficient length to provide a test length clear of the grips of at least nine times the larger cross-sectional dimension. The load is applied at a constant loading-head movement so adjusted that maximum load is reached within (300 ± 120) s. Tensile strength is given by the equation

$$f_{t,0} = \frac{F_{\text{max}}}{A} \tag{7.3}$$

where $f_{t,0}$ is the tensile strength in (N/mm^2), F_{max} is the maximum load achieved (in N), and A is the cross-sectional area (in mm^2).

COMPRESSION STRENGTH PARALLEL TO THE GRAIN

The test piece is of full cross-section throughout its length and of length six times the smaller cross-sectional dimension. Great care must be exercised to ensure that the end surfaces are plane and parallel to one another and perpendicular to the axis of the test piece. The test piece must be loaded concentrically using spherically seated loading-heads or other devices which permit the application of compressive load without inducing bending. Under test the maximum load should be reached within (300 ± 120) s.

The compressive strength is given by the equation

$$f_{\text{c},0} = \frac{F_{\text{max}}}{A} \tag{7.4}$$

where $f_{\text{c},0}$ is the compressive strength in (N/mm^2), F_{max} is the maximum load achieved (in N), and A is the cross-sectional area (in mm^2).

7.3 Strength values

7.3.1 *Derived using small clear test pieces*

For those strength properties described above in Section 7.2.2.1, with the exception of tensile strength parallel to the grain, the mean values and standard deviations (see below) are presented in Table 7.1 for a selection of timbers covering the range in densities to be found in hardwoods and softwoods. Many of the timbers whose elastic constants were presented in Table 6.1 are included. All values relate to a moisture content which is in equilibrium with

Table 7.1 Average value (upper) and standard deviation (lower) of various mechanical properties of selected timbers at 12% moisture content from small clear test pieces

	Density when dry (kg/m^3)	Static bending in three-point loading: Modulus of rupture (N/mm^2)	Modulus of elasticity (N/mm^2)	Energy to max. load ($mm\ N/mm^3$)	Energy to fracture ($mm\ N/mm^3$)	Impact: drop of hammer (m)	Compression: parallel to grain (N/mm^2)	Hardness: on side grain (N)	Shear: parallel to grain (N/mm^2)	Cleavage: Radial plane (N/mm width)	Tangential plane (N/mm width)
Hardwoods											
Balsa	176	23	3200	0.018	0.035		15.5		2.4		
		7.3	1060	0.007	0.017		4.43		0.62		
Obeche	368	54	5500	0.058	0.095	0.48	28.2	1910	7.7	9.3	8.4
		6.5	620	0.010	0.015	0.072	3.00	268	0.67	1.82	1.58
Mahogany	497	78	9000	0.070	0.128	0.58	46.4	3690	11.8	10.0	14.0
(*Khaya ivorensis*)		15.0	1520	0.026	0.044	0.149	8.45	816	2.56	2.08	2.90
Sycamore	561	99	9400	0.121	0.163	0.84	48.2	4850	17.1	16.8	27.3
		11.0	1160	0.028	0.049	0.136	4.83	639	2.32	2.95	3.91
Ash	689	116	11 900	0.182	0.281	1.07	53.3	6140	16.6		
		16.6	2170	0.045	0.097	0.216	7.73	1158	2.52		
Oak	689	97	10 100	0.093	0.167	0.84	51.6	5470	13.7	14.5	20.1
		16.8	1960	0.026	0.051	0.209	7.98	911	2.38	2.86	2.08
Afzelia	817	125	13 100	0.100	0.203	0.79	79.2	7870	16.6	10.5	13.3
		26.6	1760	0.043	0.087	0.215	12.02	914	2.28	2.00	2.49
Greenheart	977	181	21 000	0.213	0.395	1.35	89.9	10 450	20.5	17.5	22.2
		20.9	1990	0.047	0.088	0.207	8.49	1531	3.06	4.79	4.97
Softwoods											
Norway spruce	417	72	10 200	0.086	0.116	0.58	36.5	2140	9.8	8.4	9.1
(European)		10.2	2010	0.022	0.040	0.116	5.26	353	1.44	1.07	1.20
Yellow pine	433	80	8300	0.089	0.097	0.56	42.1	2050	9.3	8.2	11.6
(Canada)		10.9	1440	0.015	0.019	0.100	6.14	473	1.61	1.57	1.77
Douglas fir (UK)	497	91	10 500	0.097	0.172	0.69	48.3	3420	11.6	9.5	11.4
		16.9	2160	0.038	0.081	0.200	8.03	865	2.29	1.90	2.17
Scots pine (UK)	513	89	10 000	0.103	0.134	0.71	47.4	2980	12.7	10.3	13.0
		16.9	2130	0.032	0.053	0.167	9.25	697	2.45	1.82	2.47
Caribbean pitch	769	107	12 600	0.126	0.253	0.91	56.1	4980	14.3	12.1	13.3
pine		14.5	1800	0.042	0.060	0.196	7.76	1324	2.81	1.23	1.58

a relative humidity of 65% at 20 °C. These are of the order of 12% and the timber is referred to as 'dry'. The modulus of elasticity has also been included in Table 7.1.

Table 7.1 is compiled from data presented in Bulletin 50 of the Forest Products Research Laboratory (Lavers, 1969) which lists data for both the dry and green states for 200 species of timber. The upper line for each species provides the estimated average value; the lower line contains the standard deviation.

In Table 7.2 tensile strength parallel to the grain is listed for certain timbers and it is in this mode that timber is at its strongest.

Comparison of these values with those for compression strength parallel to the grain in Table 7.1 will indicate that, unlike many other materials, the compression strength is only about one-third that of tensile strength along the grain.

7.3.2 *Derived using structural size test pieces*

Structural strength values may still be recorded and used as mean values with a standard deviation where national standards are still being used in testing (e.g. BS 5820 (1979)) and design (e.g. BS 5268 Part 2 (1996)).

An example of mean strength values derived from testing dry structural size timber to BS 5820 (1979) is given in Table 7.3 for each of the strength modes described in Section 7.2.2.2. Not only are these mean values considerably lower than the mean values derived from small clear test pieces (Tables 7.1 and 7.2), but the tensile strengths are now lower than the compression strengths. This is directly related to the presence of knots and associated distorted grain in the structural-size test pieces.

However, by the year 2010 all structural test work and design within Europe will have to be carried out according to the new European standards by testing to EN 384 (1995) and EN 408 (1995) and designing using Eurocode 5 (ENV

Table 7.2 Tensile strength parallel to the grain of certain timbers using small clear test pieces

Timber	*Moisture content (%)*	*Tensile strength (N/mm²)*
Hardwoods		
Ash (home grown)	13	136
Beech (home grown)	13	180
Yellow poplar (imported)	15	114
Softwoods		
Scots pine (home grown)	16	92
Scots pine (imported)	15	110
Sitka spruce (imported)	15	139
Western hemlock (imported)	15	137

Table 7.3 Mean values for dry strength derived from structural size test pieces (approximately 97 × 47 mm) using BS 5820 (1979); moisture content = 15–18%

Timber	*Mean values (N/mm²)*		
	Bending	*Tension*	*Compression*
Sitka spruce (UK)	32.8	19.7	29.5
Douglas fir (UK)	35.7	21.4	32.1
Spruce/pine/fir (Canada)	43.9	26.3	39.5
Norway spruce (Baltic)	50.9	30.5	45.8

1995–1–1 (1994)). Within the European system, the *characteristic value* for the strength properties is taken as the 5 percentile value; for the modulus of elasticity there are two characteristic values, the 5 percentile value and the mean or 50 percentile value.

The sample 5 percentile value is determined for each sample by the equation

$$f_{05} = f_r \tag{7.5}$$

where f_{05} is the sample 5 percentile value, and f_r is obtained by ranking all the test values for a sample in ascending order. The 5 percentile value is the test value for which 5% of the values are lower. If this is not an actual test value (i.e. the number of test values is not divisible by 20) then interpolation between the two adjacent values is permitted.

The characteristic value of strength (f_k) is calculated from

$$f_k = \bar{f}_{05} k_s k_v \tag{7.6}$$

where $\bar{f}_{05}$ is the mean (in N/mm²) of the adjusted 5 percentile values (f_{05}) for each sample (see above) weighted according to the number of pieces in each sample, k_s is a factor to adjust for the number of samples and their size, and k_v is a factor to allow for the lower variability of f_{05} values from machine grades in comparison with visual grades; for visual grades k_v = 1.0, and for machine grades k_v = 1.12

7.4 Variability in strength values

In Chapter 1 attention was drawn to the fact that timber is a very variable material and that for many of its parameters, such as density, cell length and microfibrillar angle of the S_2 layer, distinct patterns of variation could be established within a growth ring, outwards from the pith towards the bark, upwards in the tree, and from tree to tree. The effects of this variation in structure are all too apparent when mechanical tests are performed.

An efficient estimator of the variability which occurs in any one property is the *sample standard deviation*, denoted by s. It is the square root of the variance and is derived from the formula

$$s = \sqrt{\frac{\sum x^2 - \left(\sum x\right)^2 / n}{n - 1}} \tag{7.7}$$

where x stands for every item successively and n is the number of items in the sample of either small clear, or structural size test pieces.

The frequency diagram is typical of the manner in which many strength properties are distributed, and in general it approximates to the theoretical normal distribution curve. Results for compression strength of small clear test pieces of western hemlock are presented on a frequency basis in Figure 7.4; superimposed on this is the plot of a normal distribution and the degree of fit will be seen to be very good.

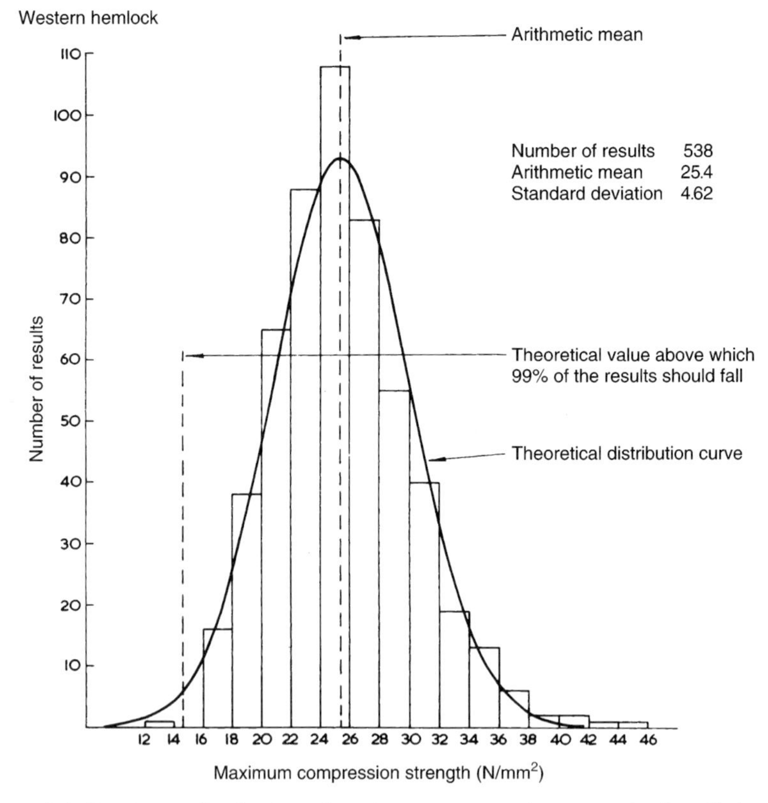

Figure 7.4 Frequency distribution of maximum compression strength of small clear test pieces of Western hemlock. (© BRE.)

In a normal distribution approximately 68% of the results should lie, in theory, within $+1 \times s$ and $-1 \times s$ of the mean, and 99.87% should fall within $\pm 3 \times s$ of the mean.

As a general rule, a normal distribution curve fits the data from small clear test pieces better than does the data from structural size test pieces.

The standard deviation provides a measure of the variability, but in itself gives little impression of the magnitude unless related to the mean. This ratio is known as the *coefficient of variation* (CV), i.e.

$$\mathrm{CV} = \frac{s}{\text{mean}} \% \qquad (7.8)$$

The coefficient of variation varies considerably, but is frequently under 15% for many biological applications. However, reference to Table 7.1 will indicate that this value is frequently exceeded. For design purposes the two most important properties are the moduli of elasticity and rupture which have coefficients of variation typically in the range 10–30%.

7.5 Interrelationships among the strength properties

7.5.1 *Modulus of rupture and modulus of elasticity*

A high correlation exists between the modulus of rupture (MOR) and the modulus of elasticity (MOE) for a particular species. Figure 7.5 shows an example of this relationship. It is doubtful whether this correlation between MOR and MOE represents any causal relationship; rather, it is more probable that the correlation arises as a result of the strong correlation that exists between density and each modulus. Whether it is a causal relationship or not, it is nevertheless put to good advantage for it forms the basis of the stress grading of timber by machine (see Section 7.8.2.5). The stiffness of a piece of timber is measured as it is deflected between rollers and this is used to predict its strength.

7.5.2 *Impact bending and total work*

Good correlations have been established between the height of drop in impact bending test and both *work to maximum load* and *total work*, in that generally the correlation is higher with the latter property.

7.5.3 *Hardness and compression perpendicular to the grain*

Correlation coefficients of 0.902 and 0.907 have been established between hardness and compression strength perpendicular to the grain of timber at 12%

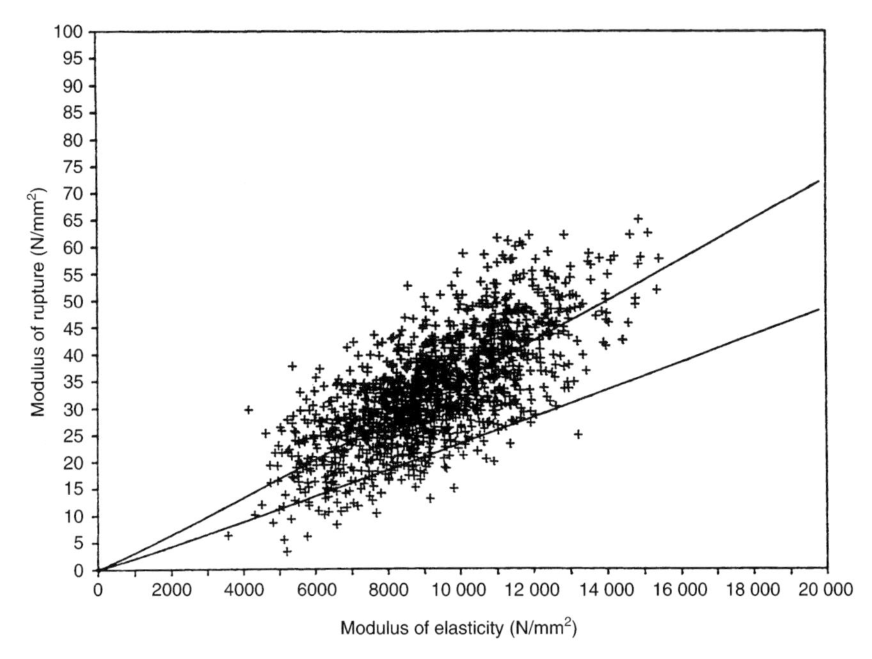

Figure 7.5 The relationship between modulus of rupture (bending strength) and modulus of elasticity for 1348 test results on UK-grown Sitka spruce, with the mean regression line and 5 percentile exclusion line superimposed. The equation of the mean regression line is MOR = 0.002065 × MOE. (© BRE.)

moisture content and timber in the green state, respectively. It is general practice to predict the compression strength from the hardness result using the equations

$$Y_{12} = 0.00147x_{12} + 1.103 \quad (7.9)$$

$$Y_g = 0.00137x_g - 0.207 \quad (7.10)$$

where Y_g and Y_{12} are compression strength perpendicular to the grain (in N/mm^2) for green timber and timber at 12% moisture content, respectively, and x_g and x_{12} are hardness (in N).

7.6 Factors affecting strength

Many of the variables noted in the previous chapter as influencing stiffness also influence the various strength properties of timber. Once again, these can be regarded as being either material dependent or manifestations of the environment.

7.6.1 *Anisotropy and grain angle*

The marked difference between the longitudinal and transverse planes in both shrinkage and stiffness has been discussed in previous chapters. Strength likewise is directionally dependent and the degree of anisotropy present in both tension and compression is presented in Table 7.4 for small clear test pieces of Douglas fir. Irrespective of moisture content, the highest degree of anisotropy is in tension (48:1). This reflects the fact that the highest strength of clear straight-grained timber is in tension along the grain, whereas the lowest is in tension perpendicular. A similar degree of anisotropy is present in the tensile stressing of both glass-reinforced plastics and carbon-fibre-reinforced plastics when the fibre is laid up in parallel strands.

Table 7.4 also demonstrates that the degree of anisotropy in compression is an order of magnitude less than in tension. Although the compression strengths are markedly affected by moisture content, tensile strength appears to be relatively insensitive. The comparison of tension and compression strengths along the grain in Table 7.4 reveals that clear straight-grained timber, unlike most other materials, has a tensile strength considerably greater than the compression strength. In structural timber containing knots and distorted grain, the opposite is the norm (see Table 7.3)

Anisotropy in strength is due in part to the cellular nature of timber and in part to the structure and orientation of the microfibrils in the wall layers. Bonding along the direction of the microfibrils is covalent, whereas bonding between microfibrils is by hydrogen bonds. Consequently, as the majority of the microfibrils are aligned at only a small angle to the longitudinal axis, it will be easier to rupture the cell wall if the load is applied perpendicular than if applied parallel to the fibre axis.

As timber is an anisotropic material, it follows that the angle at which stress is applied relative to the longitudinal axis of the cells will determine the ultimate strength of the timber. Figure 7.6 illustrates that over the range 0–45° tensile strength is much more sensitive to grain angle than is compression strength. However, at angles as high as 60° to the longitudinal axis both tension and compression strengths have fallen to only about 10% of their value in straight-grained timber. The sensitivity of strength to grain angle in clear straight-grained timber is identical with that for fibre orientation in both glass-fibre- and carbon-fibre-reinforced plastics.

Table 7.4 Anisotropy in strength in small, clear test pieces

		Tension			*Compression*		
Timber	*Moisture content (%)*	*Parallel (N/mm^2)*	*Perpendicular (N/mm^2)*	*Parallel: perpendicular*	*Parallel (N/mm^2)*	*Perpendicular (N/mm^2)*	*Parallel: perpendicular*
Douglas fir	> 27	131	2.69	48.7	24.1	4.14	5.82
Douglas fir	12	138	2.90	47.6	49.6	6.90	7.19

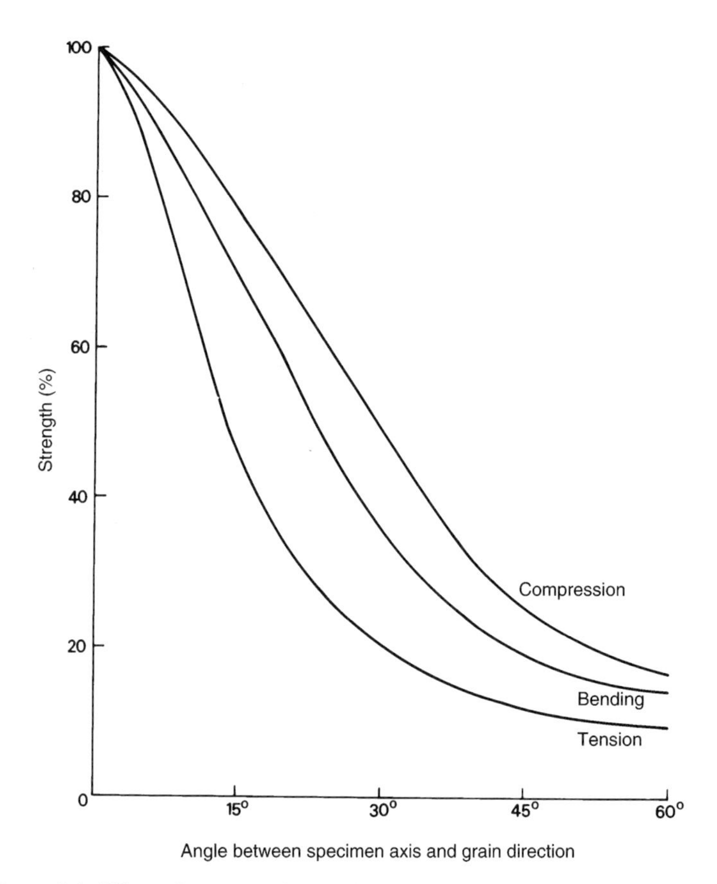

Figure 7.6 Effect of grain angle on the tensile, bending and compression strength of timber. (After R. Baumann (1922).)

It is possible to obtain an approximate value of strength at any angle to the grain from knowledge of the corresponding values for both parallel and perpendicular to the grain using the following formula which, in its original form, was credited to Hankinson:

$$f_\theta = \frac{f_L \times f_T}{f_L \sin^n \theta + f_T \cos^n \theta} \tag{7.11}$$

where f_θ is the strength property at angle θ from the fibre direction, f_L is the strength parallel to the grain, f_T is the strength perpendicular to grain, and n is an empirically determined constant; in tension, $n = 1.5$–2, in compression $n = 2$–2.5. The equation has also been used for stiffness where a value of 2 for n has been adopted. The Hankinson formula has been shown to be independent of temperature (Suzuki *et al.*, 1982).

7.6.2 Knots

Knots are associated with distortion of the grain and as even slight deviations in grain angle reduce the strength of the timber appreciably, it follows that knots will have a marked influence on strength. The significance of knots, however, will depend on their size and distribution both along the length of a piece of timber and across its section. Knots in clusters are more important than knots of a similar size which are evenly distributed, whereas knots on the top or bottom edge of a beam are more significant than those in the centre; large knots are much more critical than small knots.

It is very difficult to quantify the influence of knots. One of the parameters that has been successfully used is the *knot area ratio*, which relates the sum of the cross-sectional area of the knots at a cross-section to the cross-sectional area of the piece. The loss in bending strength that occurred with increasing knot area ratio in 200 homegrown Douglas fir boards is illustrated in Figure 7.7.

Tensile strength of structural size spruce and Douglas fir has been related to knot ratio and density; strength also increases with increasing length of test piece (Burger and Glos, 1996).

The very marked reduction in tensile strength of structural size timber compared with small clear test pieces (Tables 7.2 and 7.3) is due primarily to the presence and influence of knots in the former.

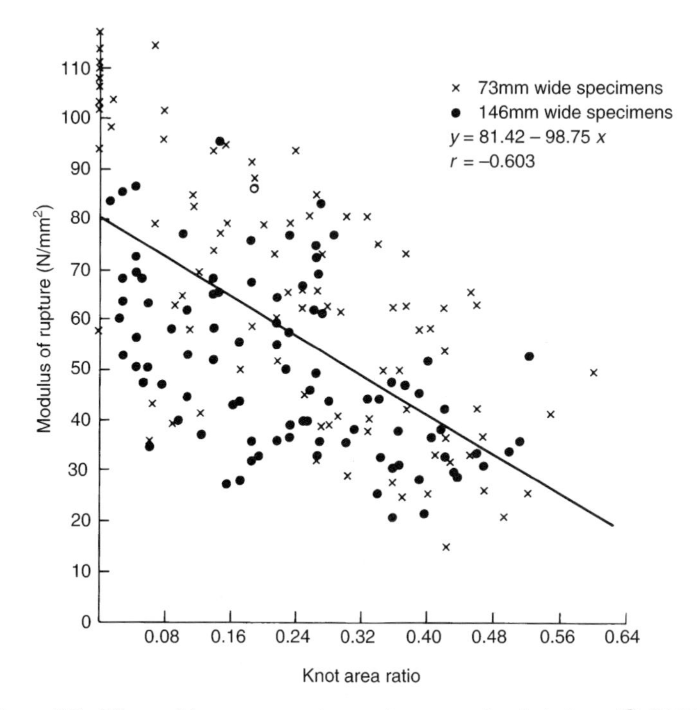

Figure 7.7 Effect of knot area ratio on the strength of timber. (© BRE.)

7.6.3 Density

In Chapters 1 and 3 density was shown to be a function of cell wall thickness and therefore dependent on the relative proportions of the various cell components and also the level of cell wall development of any one component. However, variation in density is not restricted to different species, but can occur to a considerable extent within any one species and even within a single tree. Some measure of the interspecific variation that occurs can be obtained from both Figure 3.1 and the limited amount of data in Tables 6.1 and 7.1. It will be observed from the latter that, as density increases, so stiffness and the various strength properties increase. Density continues to be the best prediction of timber strength as high correlations between strength and density are a common feature in timber studies.

Most of the relationships that have been established throughout the world between the various strength properties and timber density take the form

$$f = kg^n \tag{7.12}$$

where f is any strength property, g is the specific gravity, k is a proportionality constant differing for each strength property, and n is an exponent that defines the shape of the curve. An example of the use of this expression on the results of over 200 species tested in compression parallel to the grain is presented in Figure 7.8. The correlation coefficient between compression strength and density of the timber at 12% moisture content was 0.902.

Similar relationships have been found to hold for other strength properties, although in some the degree of correlation is considerably lower. This is the case in tension parallel to the grain where the ultrastructure probably plays a more significant role.

Over the range of density of most of the timbers used commercially, the relationship between density and strength can safely be assumed to be linear with the possible exception of shear and cleavage; similarly, within a single species, the range is low and the relationship can again be treated as linear.

7.6.4 Ring width

As density is influenced by the rate of growth of the tree it follows that variations in ring width will change the density of the timber and hence the strength. However, the relationship is considerably more complex than it first appears. In the ring-porous timbers such as oak and ash (see Chapter 1), increasing rate of growth (ring width) results in an increase in the percentage of the latewood which contains most of the thick-walled fibres; consequently, density will increase and so will strength. However, there is an upper limit to ring width beyond which density begins to fall owing to the inability of the tree to produce the requisite thickness of wall in every cell.

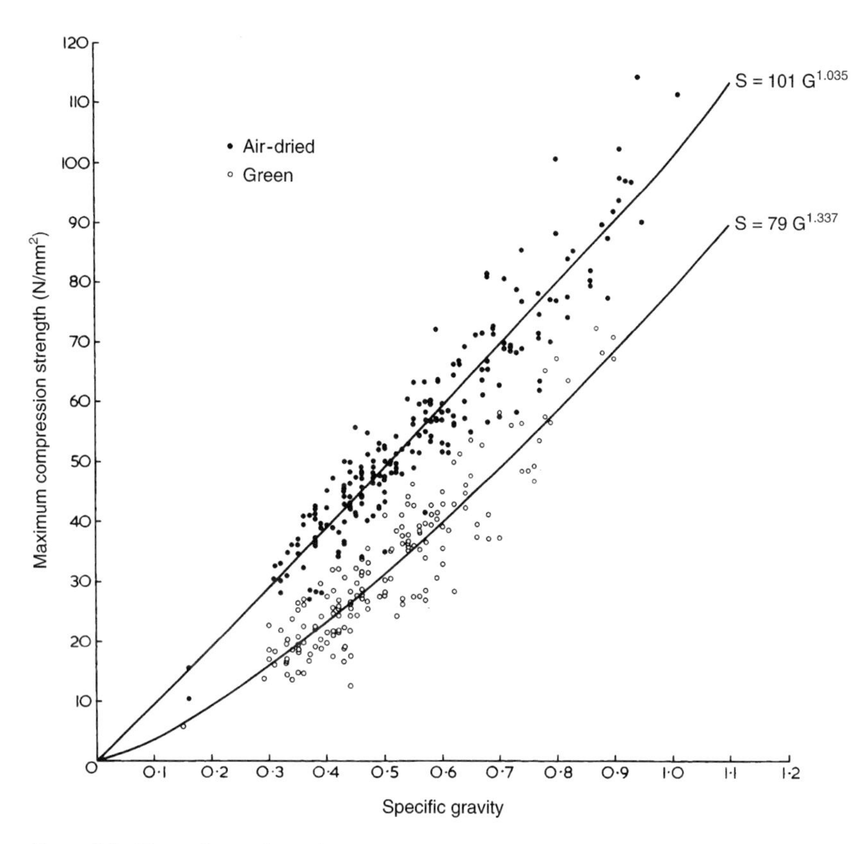

Figure 7.8 The relationship of maximum compression strength to specific gravity for 200 species tested in the green and dry states. (© BRE.)

In the diffuse-porous timbers such as beech, birch and khaya, where there is uniformity in structure across the growth ring, increasing rate of growth (ring width) has no effect on density unless, as before, the rate of growth is excessive.

In the softwoods, however, increasing rate of growth results in an increased percentage of the low-density earlywood and consequently both density and strength decrease as ring width increases. Exceptionally, it is found that very narrow rings can also have very low density. This is characteristic of softwoods from the very northern latitudes where latewood development is restricted by the short summer period. Hence ring width of itself does not affect the strength of the timber; nevertheless, it has a most important indirect effect working through density.

7.6.5 *Ratio of latewood to earlywood*

As the latewood comprises cells with thicker walls, it follows that increasing the percentage of latewood will increase the density and therefore the strength of the timber. Differences in strength of 150–300% between the late and early-wood are generally explained in terms of the thicker cell walls of the former; however, some workers maintain that when the strengths are expressed in terms of the cross-sectional area of the cell wall the latewood cell is still stronger than the earlywood. Various theories have been advanced to account for the higher strength of the latewood wall material. The more acceptable are couched in terms of the differences in microfibrillar angle in the middle layer of the secondary wall, differences in degree of crystallinity and, lastly, differences in the proportion of the chemical constituents.

7.6.6 *Cell length*

As the cells overlap one another, it follows that there must be a minimum cell length below which there is insufficient overlap to permit the transfer of stress without failure in shear occurring. Some investigators have gone further and have argued that there must be a high degree of correlation between the length of the cell and the strength of cell wall material, because a fibre with high strength per unit of cross-sectional area would require a larger area of overlap in order to keep constant the overall efficiency of the system.

7.6.7 *Microfibrillar angle*

The angle of the microfibrils in the S_2 layer has a most significant effect in determining the strength of wood. Figure 7.9 illustrates the marked reduction in tensile strength that occurs with increasing angle of the microfibrils; the effect on strength closely parallels that which occurs with changing grain angle.

7.6.8 *Chemical composition*

In Chapter 1 the structure of the cellulose molecule was described and emphasis was placed on the existence in the longitudinal plane of covalent bonds both within the glucose units and also linking them together to form filaments containing from 5000 to 10 000 units. There is little doubt that the high tensile strength of timber owes much to the existence of this covalent bonding. Certainly, experiments in which many of the β-1, 4 linkages have been ruptured by gamma irradiation resulting in a decrease in the number of glucose units in the molecule from over 5000 to about 200 have resulted in a most marked reduction in tensile strength. It has also been shown that timber with inherently low molecular lengths, such as compression wood, has a lower than normal tensile strength.

Until the 1970s it had been assumed that the hemicelluloses that constitute about half of the matrix material played little or no part in determining the strength of timber. However, it has now been demonstrated that some of the hemicelluloses are orientated within the cell wall and it is now thought that these will be load bearing.

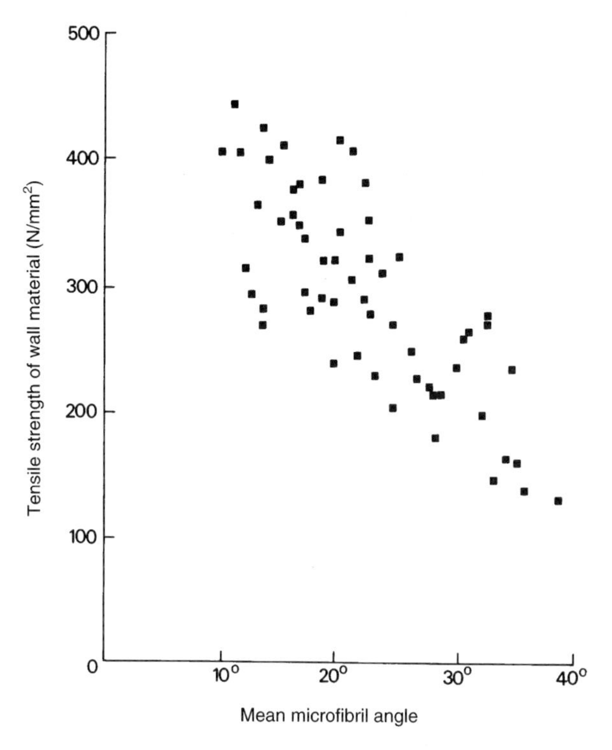

Figure 7.9 Effect of microfibrillar angle on the tensile strength of *Pinus radiata* blocks. (From I. D. Cave (1969) *Wood Science and Technology*, **3**, 40–48, reproduced by permission of Springer-Verlag.)

It is known that lignin is less hydrophilic than either the cellulose or hemicelluloses and, as indicated earlier, at least part of its function is to protect the more hydrophilic substances from the ingress of water and consequent reduction in strength. Apart from this indirect effect on strength, lignin is thought to make a not too insignificant direct contribution. Much of the lignin in the cell wall is located in the primary wall and in the middle lamella. As the tensile strength of a composite with fibres of a definite length will depend on the efficiency of the transfer of stress by shear from one fibre to the next, it will be appreciated that in timber the lignin is playing a most important role. Compression strength along the grain has been shown to be affected by the degree of lignification not between the cells, but rather within the cell wall, when all the other variables have been held constant.

It would appear, therefore, that both the fibre and the matrix components of the timber composite are contributing to its strength as in fact they do in most composites; the relative significance of the fibre and matrix roles will vary with the mode of stressing.

7.6.9 Reaction wood

7.6.9.1 Compression wood

The chemical and anatomical properties of this abnormal wood, which is found only in the softwoods, were described in Chapter 1. When stressed, it is found that the tensile strength and toughness is lower and the compressive strength higher than that of normal timber. Such differences can be explained in terms of the changes in fine structure and chemical composition.

7.6.9.2 Tension wood

This second form of abnormal wood, which is found only in the hardwoods, has tensile strengths higher and compression strengths lower than normal wood; again, this can be related to changes in fine structure and chemical composition (see Chapter 1).

7.6.10 Moisture content

The marked increase in strength on drying from the fibre saturation point to oven-dry conditions was described in detail in Chapter 4 and illustrated in Figure 4.2. Experimentation has indicated the probability that at moisture contents of less than 2% the strength of timber may show a slight decrease rather than the previously accepted continuation of the upward trend.

Confirmatory evidence of the significance of moisture content on strength is forthcoming from Figure 7.8 in which the regression line for over 200 species of compression strength of green timber against density is lower than that for timber at 12% moisture content; strength data for timber are generally presented for these two levels of moisture content (Lavers, 1969).

However, reference to Table 7.4 will indicate that the level of moisture has almost no effect on the tensile strength parallel to the grain. This strength property is determined by the strength of the covalent bonding along the molecule and, as the crystalline core is unaffected by moisture (Chapter 1), retention of tensile strength parallel to the grain with increasing moisture content is to be expected.

Within certain limits and excluding tensile strength parallel, the regression of strength, expressed on a logarithmic basis, and moisture content can be plotted as a straight line. The relationship can be expressed mathematically as

$$\log_{10} f = \log_{10} f_s + k(\mu_s - \mu) \tag{7.13}$$

where f is the strength at moisture content u, f_s is the strength at the fibre saturation point, μ_s is the moisture content at the fibre saturation point, and k is a constant. It is possible, therefore, to calculate the strength at any moisture content below the fibre saturation point, assuming f_s to be the strength of the green timber and μ_s to be 25%. This formula can also be used to determine the strength changes that occur for a 1% increase in moisture content over certain ranges. Table 7.5 illustrates, for small clear test pieces, how the change in strength per unit change in moisture content is non-linear.

This relationship between moisture content and strength may not always apply when the timber contains defects as is the case with structural size timber. Thus, it has been shown that the effect of moisture content on strength diminishes as the size of knots increase.

The relationship between moisture content and strength presented above, even for knot-free timber, does not always hold for the impact resistance of timber. In some timbers, though certainly not all, impact resistance or toughness of green timber is considerably higher than it is in the dry state; the impact resistances of green ash, cricket bat willow and teak are approximately 10%, 30% and 50% higher, respectively, than the values at 12% moisture content. This increase could be related to the increased flexibility (decreased stiffness) in the wet state but, as the latter is common to all timbers, it is difficult to understand why the toughness of green timber is not higher than dry timber for all species. Certainly it is not possible to account for the disparity in behaviour in terms of differences in microscopic structure.

In the case of structural timber several types of model have been proposed to represent moisture–property relationships (e.g. Green *et al.*, 1986; Madsen, 1982; Barrett and Lau, 1991). Unlike earlier models, that by Barrett and Lau utilises a linear surface model to represent moisture content effects on bending strength and bending capacity, making the assumption that strength is linearly related to moisture content below the fibre saturation point. Such a linear model is capable of incorporating the fact that increases in strength with drying are greater for high-strength structural timber than for low-strength material. The model fitted the experimental data nearly as well as the earlier quadratic surface models, but is easier to use.

Table 7.5 Percentage change in strength and stiffness of Scots pine timber per 1% change in moisture content (Lavers, 1983)

Property	*Moisture range (%)*		
	6–10	*12–16*	*20–24*
Modulus of elasticity (MOE: stiffness)	0.21	0.18	0.15
Modulus of rupture (MOR: bending strength)	4.2	3.3	2.4
Compression, perpendicular to the grain	2.7	2.0	1.40
Hardness	0.058	0.053	0.045
Shear, parallel to the grain	0.70	0.53	0.36

7.6.11 Temperature

At temperatures within the range +200 °C to –200 °C and at constant moisture content, strength properties are linearly (or almost linearly) related to temperature, decreasing with increasing temperature. However, a distinction must be made between short- and long-term effects.

When timber is exposed for short periods of time to temperatures below 95 °C the changes in strength with temperature are reversible. These reversible effects can be explained in terms of the increased molecular motions and greater lattice spacing at higher temperatures. Figure 7.10 illustrates the increase in strength that occurred in three timbers loaded in compression along the grain as temperature was progressively reduced. For all of the strength properties, with the possible exception of tensile parallel to the grain (Gerhards, 1982), a good rule of thumb is that an increase in temperature of 1°C produces a 1% reduction in their ultimate values.

At temperatures above 95 °C, or at temperatures above 65 °C for very long periods of time, there is an irreversible effect of temperature due to thermal degradation of the wood substance, generally taking the form of a marked shortening of the length of the cellulose molecules and chemical changes within the hemicelluloses (see Section 8.2.3). All strength properties show a marked

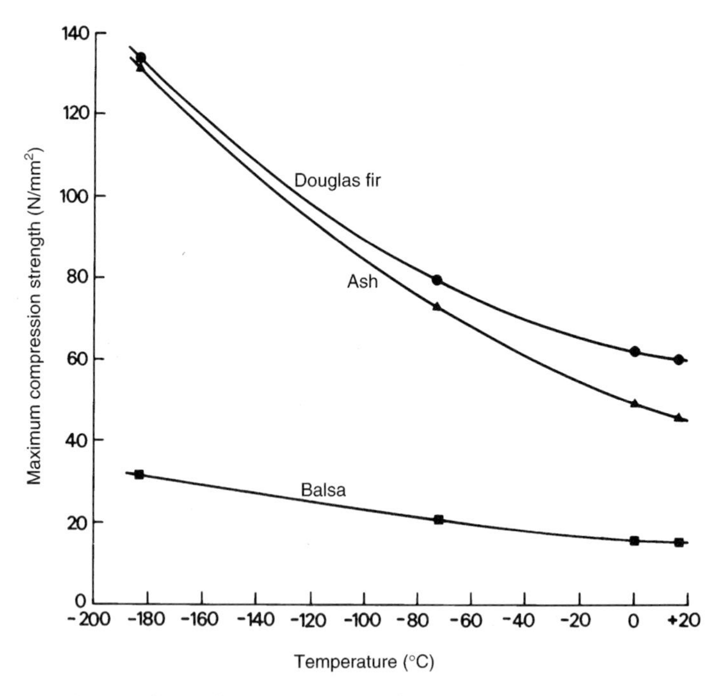

Figure 7.10 The effect of temperature on the maximum compression strength parallel to the grain in two hardwood and one softwood timbers. (© BRE.)

reduction with temperature, but toughness is particularly sensitive to thermal degrade. Repeated exposure to elevated temperature has a cumulative effect and usually the reduction is greater in the hardwoods than in the softwoods. Even exposure to cyclic changes in temperature over long periods of time has been shown to result in thermal degradation and loss in strength and especially toughness (Moore, 1984).

The effect of temperature is very dependent on moisture content, sensitivity of strength to temperature increasing appreciably as moisture content increases (Figure 7.11), as occurs also with stiffness (Figure 6.14); these early results have been confirmed by Gerhards (1982). The relationship between strength, moisture content and temperature appear to be slightly curvilinear over the range 8–20% moisture content and –20°C to 60°C. However, in the case of toughness, whereas at low moisture content it is found that toughness decreases with increasing temperature, at high moisture contents toughness actually increases with increasing temperature.

7.6.12 *Time*

In Chapter 6 timber was described as a viscoelastic material and as such its mechanical behaviour will be time dependent. Such dependence will be apparent in terms of its sensitivity to both rate of loading and duration of loading.

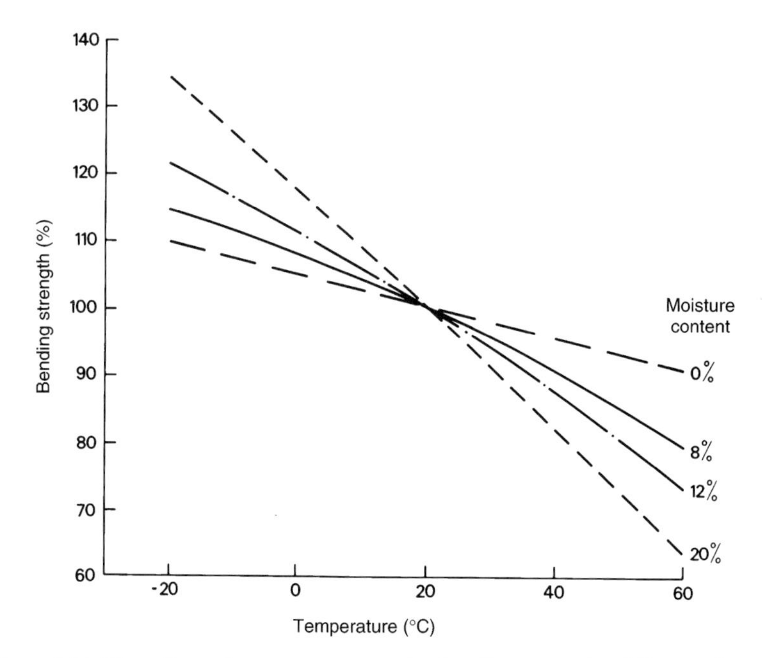

Figure 7.11 The effect of temperature on the bending strength of *Pinus radiata* timber at different moisture contents.

7.6.12.1 Rate of loading

Increase in the rate of loading (ROL) results in increased strength values, the increase in 'green' timber being some 50% greater than that of timber at 12% moisture content; however, strain to failure actually decreases. A variety of explanations have been presented to account for this phenomenon, most of which are based on the theory that timber fails when a critical strain has been reached and consequently at lower rates of loading viscous flow or creep is able to occur resulting in failure at lower loads.

The various standard testing procedures adopted throughout the world set tight limits on the speed of loading in the various tests. Unfortunately, such recommended speeds vary throughout the world, thereby introducing errors in the comparison of results from different laboratories; the introduction of European standards (ENs) and the wider use of international standards (ISOs), should give rise to greater uniformity in the future.

7.6.12.2 Duration of load

In terms of the practical use of timber, the duration of time over which the load is applied is perhaps the single most important variable. Many investigators have worked in this field and each has recorded a direct relationship between the length of time over which a load can be supported at constant temperature and moisture content and the magnitude of the load. This relationship appears to hold true for all loading modes, but is especially important for bending strength.

The modulus of rupture (maximum bending strength) will decrease in proportion, or nearly in proportion, to the logarithm of the time over which the load is applied; failure in this particular time-dependent mode is termed *creep rupture* or *static fatigue*. Wood (1951) indicated that the relationship was slightly different for ramp and constant loading, was slightly curvilinear and that there was a distinct levelling off at loads approaching 20% of the ultimate short-term strength such that a critical load or stress level occurs below which failure is unlikely to occur. The hyperbolic curve that fitted Wood's data best for both ramp and sustained loading and which became known as the *Madison curve* is illustrated in Figure 7.12.

Other workers have reported a linear relationship, though a tendency to non-linear behaviour at very high stress levels has been recorded by some of them. Pearson (1972), in reviewing previous work in the field of duration of load and bending strength, plotted on a single graph the results obtained over a 30 year period and found that despite differences in method of loading (ramp or constant), species, specimen size, moisture content, or whether the timber was solid or laminated, the results showed remarkably little scatter from a straight line described by the regression

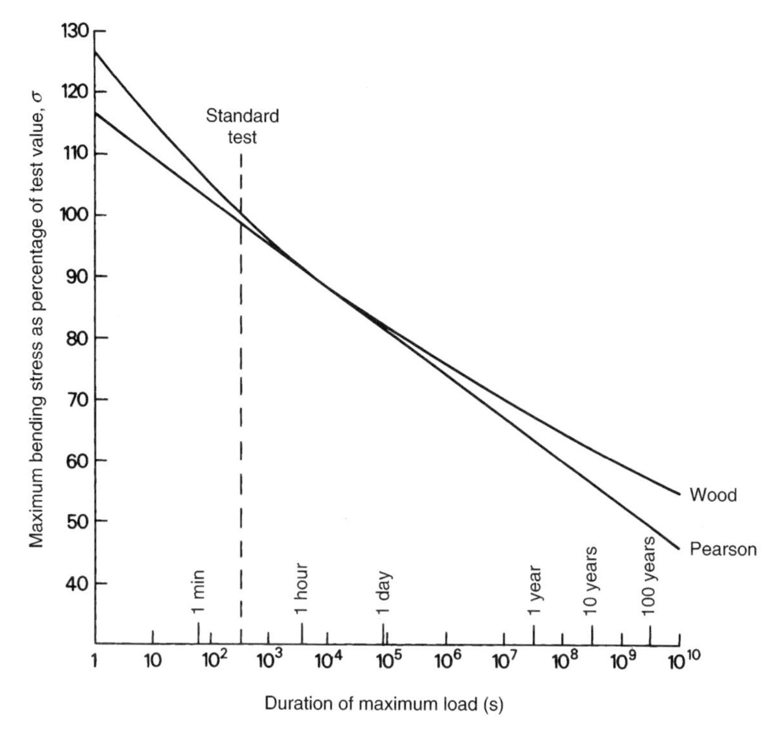

Figure 7.12 The effect of duration of load on the bending strength of timber. (After L.W. Wood (1951) *Report No. 1916*, Forest Products Laboratory, also R.G. Pearson (1972) *Holzforschung*, **26** (4), Editor: G. Stegman, Walter de Gruyter & Co., Berlin, New York.)

$$f = 91.5 - 7 \log_{10} t \tag{7.14}$$

where f is the stress level (%), and t is the effective duration of maximum load in hours. This regression, which is only slightly different from that of Clouser (1959) to which reference was made in Chapter 6, is also plotted in Figure 7.12 for comparison with the curvilinear line. Pearson's findings certainly threw doubt on the existence of a critical stress level below which creep rupture does not occur. These regressions indicate that timber beams which have to withstand a dead load for 50 years can be stressed to only 50% of their ultimate short-term strength.

Although this type of log–linear relationship is still employed for the derivation of duration of load factors for wood-based panel products, as witness the new European test method ENV 1156 (1999), this is certainly not the case with solid timber.

By the early 1970s, there was abundant evidence to indicate that the creep rupture response of structural timber beams differed considerably from the classic case for small clear test pieces described above. Madsen (1973), by developing a set of new techniques which have had a profound effect on all subsequent research in the field of duration of load (DOL), found that the DOL effect varied with timber quality (strength). Low-strength structural timber not only had less DOL effect than high strength timber, but it also had a DOL effect that was significantly smaller than that predicted from the Madison curve which is based on small clear test pieces. In further tests (Madsen and Barrett, 1976; Madsen, 1979), it was discovered that at high stress ratios the DOL effect is less severe for large sized timber than would be predicted from the Madison curve.

Between 1970 and 1985 an extensive amount of research was carried out in America, Canada and Europe. The historical development of the new concepts in DOL is covered in the comprehensive reviews by Tang (1994) and Barrett (1996). This research confirmed that the DOL effect in structural timber was different to that in small clear test pieces and was also less severe than the Madison curve predicted for loading periods up to 1 year. It also confirmed that high-strength timber possessed a larger DOL effect than low-strength timber.

The above test work clearly indicated that the DOL factors then in current use were conservative, and in order to obtain a more realistic prediction of time to failure, attention moved to the possible application of reliability-based design principles for the assessment of the reliability of timber members under *in-service* loading conditions. In particular, this approach led to the adoption of the concept of *damage accumulation*.

It should be appreciated that in the application of this concept there does not exist any method for quantifying the actual damage; the development of damage is simply deduced from the time-to-failure data from long-term loading experiments under a given loading history. These models generally use the stress-level history as the main variable and are thus independent of material strength. In order to calculate time to failure for a given stress history under constant temperature and humidity, the damage rate is integrated from an assumed initial value of 0 to the failure value of 1 (Morlier *et al.* 1994).

Several types of damage accumulation models have been recorded, of which the most important are those listed by Morlier *et al.* (1994) and Tang (1994):

- Derived from the hyperbolic Madison curve

$$\mathrm{d}\alpha/\mathrm{d}t = A(\tau - \tau_0)^B \tag{7.15}$$

- Barrett and Foschi model 1 (Barrett and Foschi, 1978)

$$\begin{aligned} \mathrm{d}\alpha/\mathrm{d}t &= A(\tau - \tau_0)^B \alpha^C \qquad && \text{if } \tau > \tau_0 \\ &= 0 && \text{if } \tau \leq \tau_0 \end{aligned} \tag{7.16}$$

- Barrett and Foschi model 2 (Barrett and Foschi, 1978)

$$\begin{aligned} \mathrm{d}\alpha/\mathrm{dt} &= A(\tau - \tau_0)^B + C\alpha \qquad && \text{if } \tau > \tau_0 \\ &= 0 && \text{if } \tau \leq \tau_0 \end{aligned} \tag{7.17}$$

- Gerhards model (Gerhards, 1979; Gerhards and Link, 1987)

$$d\alpha/dt = \exp(-A + B\tau) \quad (7.18)$$

- Foschi and Yao model (Foschi and Yao, 1986a)

$$d\alpha/dt = A(\sigma - \tau_0\sigma_U)^B + C(\sigma - \tau_0\sigma_U)^D\alpha \quad (7.19)$$

In all models α is the damage state variable (0 in undamaged state and 1 at failure), τ is a stress ratio defined as the applied stress σ divided by the short-term ultimate strength σ_U (i.e. $\tau = \sigma/\sigma_U$); τ_0 is a stress threshold below which damage is assumed not to accumulate, and A, B, C and D are model parameters.

The dependence of these damage models on the stress ratio τ results in a logistical problem, because both the short-term and the long-term strength has to be known for the same structural test piece, but the test piece can be tested only once. This problem is usually resolved by using two side-matched test pieces and assuming equal strengths! A comparison among four of the above damage-accumulation models, together with one model based on strain energy (Fridley *et al.*, 1992a – see later), is presented in Figure 7.13, from which it will be noted how large is the variability among them.

The level of both moisture content and temperature has a marked effect on time to failure. Thus, increasing levels of relative humidity result in reduced times to failure when stressed at the same stress ratios (Fridley *et al.*, 1991),

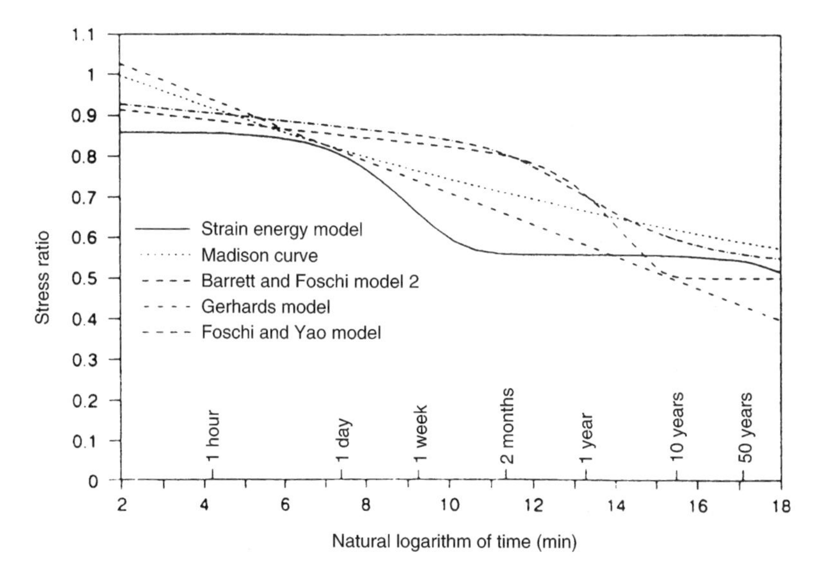

Figure 7.13 Comparison of duration of load predictions among four damage accumulation models and one model based on strain energy (From K. J. Fridley, R.C. Tang and L.A. Soltis (1992a) *J. Struct. Eng., Structural Div., ACSE* **118** (9), 2351–2369. Reproduced by permission of the publisher American Society of Civil Engineers.)

whereas varying levels of humidity have an even greater effect in reducing times to failure (Hoffmeyer, 1990; Fridley *et al.*, 1992b). It is interesting to note in passing that comparable effects of cyclic humidity have been recorded on small clear test pieces (see Schniewind (1967) and also Section 6.3.1.7 and Figure 6.21). Reduced times to failure have also been recorded for increasing temperature (Fridley *et al.*, 1989).

Few attempts have been made to apply damage accumulation models to variable humidity conditions; even those models which have been published (for example Fridley *et al.*, 1992b,c; Toratti, 1992) appear to be limited in their application and not to be able to cover the generalised situation.

As Morlier *et al.* (1994) have pointed out, the use of damage accumulation models is only one of four theoretical approaches to the question of time to failure (DOL). These phenomenological models are applicable at the macroscopic level to represent one particular type of loading arrangement.

A second approach employs fracture mechanics to study the process of slow crack growth through timber. Pitched at the microscopic level, this approach provides some complementary information on cause and effect in the failure process. Fracture mechanics has been successfully applied to time to failure studies in timber (Schniewind, 1977; Johns and Madsen, 1982; Nielsen, 1986 – as one example of the application of the *damaged viscoelastic material theory*). A full account of the various applications of fracture mechanics to timber is given in Section 7.7.2.1.

A third approach, which is again phenomenological, is the use of chemical kinetic theory (deformation kinetics) which at the molecular level relates time to failure to the breakage of bonds and their movement to more stable configurations (Caulfield, 1985; van der Put, 1983, 1993). The application of this concept to the modelling of creep is described in Section 6.3.1.5.

The fourth and last approach to duration of load is a probabilistic one and is particularly relevant to the behaviour of timber in actual structures. Such an approach is justified in terms of the high variability in timber strength; it generally uses stochastic analysis (Foschi and Yao, 1986(b)).

FAILURE THEORIES FOR TIME-DEPENDENT FRACTURE

Timber has been shown to be a viscoelastic material and in the formulation of a failure theory to account for creep rupture (static fatigue) the general theory of viscoelasticity has been adopted using some particular boundary criterion which is defined in appropriate rheological terms.

Generally, a *critical strain criteria* has been adopted; thus Ylinen (1957) combined a viscoelastic model, namely Maxwell's differential equation for relaxation, with St Venant's fracture criteria. This model was shown to agree well with data in the literature.

A similar approach has utilised the combination in series of a spring and a Maxwell element and applied the Coulomb hypothesis of a critical elastic strain.

This provided the time of onset of plastic flow which in turn was used to calculate the long-term load of the timber.

James (1962), in questioning the general applicability of Ylinen's results, suggested replacement of the linear model by a non-linear one. Even so, this critical strain approach is not sufficiently general because the total strain to rupture is dependent on the time history of loading.

An entirely different approach was the use of some energy criterion of fracture as it had been demonstrated that the total mechanical work to fracture may be constant when uniform rates of loading are adopted in the various excitation modes. However, this approach was not very fruitful in the early stages.

In the post-1970s period the classical failure theories described above have almost disappeared, to be replaced by the newer concepts of damage accumulation, fracture mechanics, chemical kinetics, and possibly to a lesser extent by stochastic analysis. Classical failure theory, however, has not disappeared completely. Thus, several energy-based criteria for failure have recently been proposed, one of which is the total strain energy criterion. This was used by Fridley *et al.* (1992a) to model the initiation of impending failure which embraces the commencement of tertiary creep in a constant load test, and the proportional limit in a ramp load test. An example of the efficacy of this model in predicting time to failure of timber (North American Select Structural grade of timber) under constant load and various cyclic climates is given in Figure 7.14.

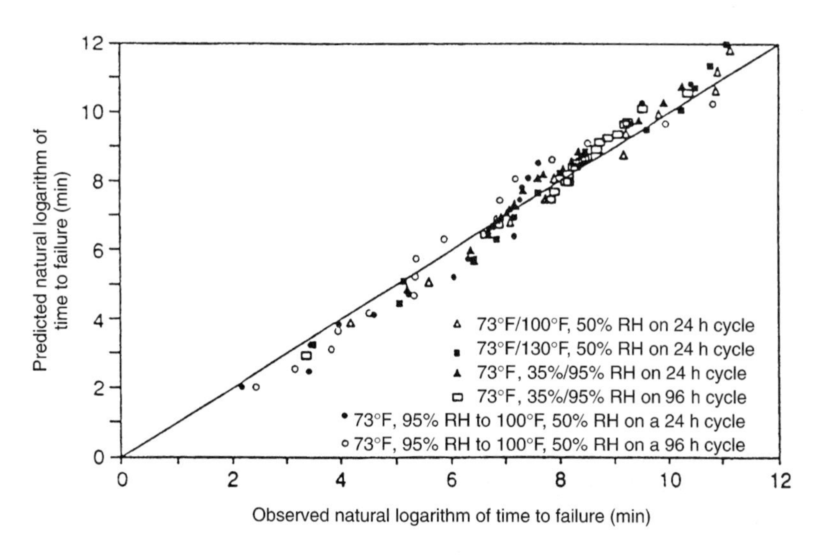

Figure 7.14 Predicted (strain energy model) versus observed time-to-failure for N. American Select Structural timber subjected to constant load and various cyclic climates. (From K.J. Fridley, R.C. Tang and L.A. Soltis (1992a) *J. Struct. Eng., Structural Div., ACSE* **118** (9), 2351–2369. Reproduced by permission of the publisher American Society of Civil Engineers.)

7.7 Strength, toughness, failure and fracture morphology

There are two fundamentally different approaches to the concept of strength and failure. The first is the classical strength of materials approach, attempting to understand strength and failure of timber in terms of the strength and arrangement of the molecules, the fibrils, and the cells by thinking in terms of a theoretical strength and attempting to identify the reasons why the theory is never satisfied.

The second and more recent approach is much more practical in concept because it considers timber in its present state, ignoring its theoretical strength and its microstructure and stating that its performance will be determined solely by the presence of some defect, however small, which will initiate on stressing a small crack. The ultimate strength of the material will depend on the propagation of this crack. Many theories have required considerable modification for their application to the different fracture modes in an anisotropic material such as timber.

Both approaches are discussed below for the more important modes of stressing.

7.7.1 *Classical approach*

7.7.1.1 *Tensile strength parallel to the grain*

Over the years a number of models have been employed in an attempt to quantify the theoretical tensile strength of timber. In these models it is assumed that the lignin and hemicelluloses make no contribution to the strength of the timber; in the light of recent investigations, however, this may be no longer valid for some of the hemicelluloses. One of the earliest attempts modelled timber as comprising a series of endless chain molecules, and strengths of the order of 8000 N/mm^2 were obtained. More recent modelling has taken into account the finite length of the cellulose molecules and the presence of amorphous regions. Calculations have shown that the stress to cause chain slippage is generally considerably greater than that to cause chain scission, irrespective of whether the latter is calculated on the basis of potential energy function or intrachain bond energies between links; preferential breakage of the cellulose chain is thought to occur at the C–O–C linkage. These important findings have led to the derivation of minimum tensile stresses of the order of 1000–7000 N/mm^2 (Mark, 1967).

The ultimate tensile strength of timber is of the order of 100 N/mm^2, though this varies considerably between species. This figure corresponds to a value between 0.1 and 0.015 of the theoretical strength of the cellulose fraction. As this accounts for only half the mass of the timber (Table 1.2) and as it is assumed, perhaps incorrectly, that the matrix does not contribute to the strength, it can be said that the actual strength of timber lies between 0.2 and 0.03 of its theoretical strength.

In attempting to integrate these views of molecular strength with the overall concept of failure it is necessary to examine strength at the next order of magnitude, namely the individual cells. It is possible to separate these by dissolution of the lignin–pectin complex cementing them together (Chapter 1 and Figures 1.7 and 1.8). Using specially developed techniques of mounting and stressing, it is possible to determine their tensile strengths. Much of this work has been done on softwood tracheids, and mean strengths of the order of 500 N/mm^2 have been recorded by a number of investigators. The strengths of the latewood cells can be up to three times that of the corresponding earlywood cells.

Individual tracheid strength is therefore approximately five times greater than that for solid timber. Softwood timber also contains parenchyma cells which are found principally in the rays, and lining the resin canals, and which are inherently weak. Many of the tracheids tend to be imperfectly aligned and there are numerous discontinuities along the cell; consequently it is to be expected that the strength of timber is lower than that of the individual tracheids. Nevertheless, the difference is certainly substantial and it seems doubtful if the features listed above can account for the total loss in strength, especially when it is realised that the cells rupture on stressing and do not slip past one another.

When timber is stressed in tension along the grain, failure occurs catastrophically with little or no plastic deformation (Figure 6.2) at strains of about 1%. Visual examination of the sample usually reveals an interlocking type of fracture which can be confirmed by optical microscopy. However, as illustrated in Figure 7.15, the degree of interlocking is considerably greater in the latewood than in the earlywood. Whereas in the latewood, the fracture plane is essentially vertical, in the earlywood the fracture plane follows a series of shallow zigzags in a general transverse plane; it is now thought that these thin-walled cells contribute very little to the tensile strength of the timber. Thus, failure in the stronger latewood region is by shear, whereas in the earlywood, although there is some evidence of shear failure, most of the rupture appears to be transwall or brittle.

Examination of the fracture surfaces of the latewood cells by electron microscopy reveals that the plane of fracture occurs either within the S_1 layer or, as is more common, between the S_1 and S_2 layers. As shear strengths are lower than tensile strengths, these observations are in accord with comments made previously on the relative superiority of the tensile strengths of individual fibres compared with the tensile strength of timber. By failing in shear this implies that the shear strength of the wall layers is lower than the shear strength of the lignin–pectin material cementing together the individual cells.

Confirmation of these views is forthcoming from the work of Mark (1967) who has calculated the theoretical strengths of the various cell wall layers and has shown that the direction and level of shear stress in the various wall layers was such as to initiate failure between the S_1 and S_2 layers. Mark's treatise received a certain amount of criticism on the grounds that he treated one cell in isolation, opening it up longitudinally in his model to treat it as a

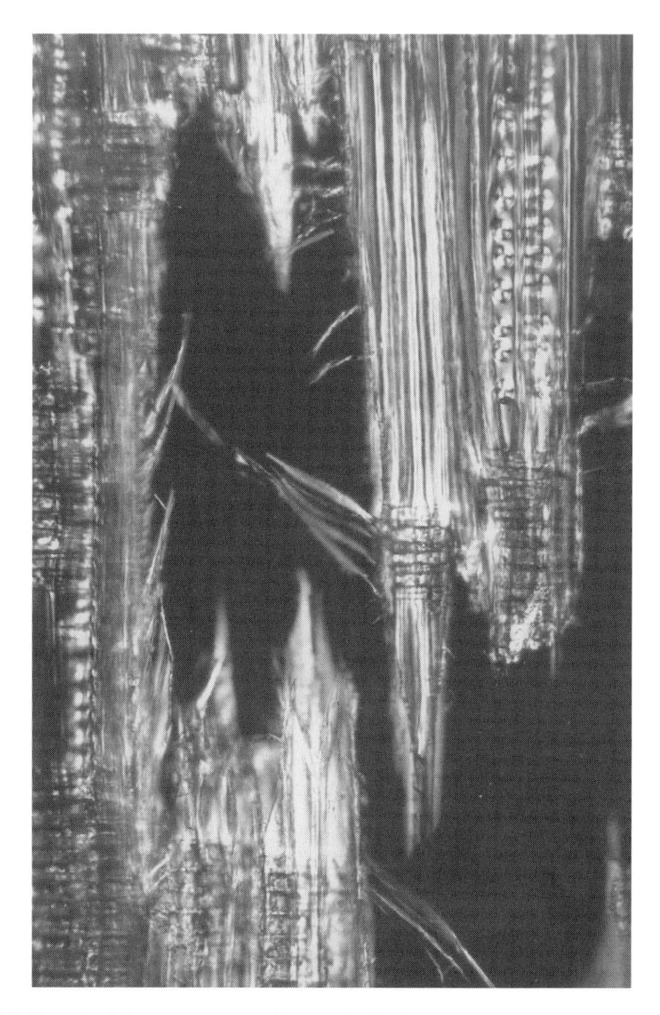

Figure 7.15 Tensile failure in spruce (*Picea abies*) showing mainly transverse cross-wall failure of the earlywood (left) and longitudinal intrawall shear failure of the latewood cells (right) (× 110, polarised light). (© BRE.)

two-dimensional structure. Nevertheless, the work marked the beginning of a new phase of investigation into elasticity and fracture, and the approach has been modified and subsequently developed. The extension of the work has explained the initiation of failure at the S_1–S_2 boundary, or within the S_1 layer, in terms of either buckling instability of the microfibrils, or the formation of ruptures in the matrix or framework giving rise to a redistribution of stress.

Thus, both the microscopic observations and the developed theories appear to agree that failure of timber under longitudinal tensile stressing is basically by shear, unless the density is low when transwall failure occurs. However, under certain conditions the pattern of tensile failure may be abnormal. First, at temperatures in excess of 100 °C, the lignin component is softened and its shear strength is reduced. Consequently, on stressing, failure will occur within the cementing material rather than within the cell wall.

Second, transwall failure has been recorded in weathering studies where the mode of failure changed from shear to brittle as degradation progressed. This was interpreted as being caused by a breakdown of the lignin and degradation of the cellulose, both of which processes would be reflected in a marked reduction in density (Turkulin and Sell, 1997).

Finally, in timber that has been highly stressed in compression before being pulled in tension, it will be found that tensile rupture will occur along the line of compression damage which, as will be explained below, runs transversely. Consequently, failure in tension is horizontal giving rise to a brittle type of fracture (see Figure 8.1).

In the literature a wide range of tensile failure criteria is recorded, the most commonly applied being some critical strain parameter, an approach supported by a considerable volume of evidence, though its lack of universal application has been pointed out by several workers. The situation is therefore very similar to that described for the criteria used to explain failure under sustained load in small clear test pieces (Section 7.6.12.2).

7.7.1.2 Compression strength parallel to the grain

Of the few attempts have been made to derive a mathematical model for the compressive strengths of timber, one of the most successful is that by Easterling *et al.* (1982). In modelling the axial and transverse compressive strength of balsa, these authors found that their theory – which related the axial strength linearly to the ratio of the wood density to the dry cell wall material density, and the transverse strength to the square of this ratio – was well supported by experimental evidence. It also appears that their simple theory for balsa may be applicable to timber of higher density.

Compression failure is a slow yielding process in which there is a progressive development of structural change. The initial stage of this sequence appears to occur at a stress of about 25% of the ultimate failing stress (Dinwoodie, 1968), though Keith (1971) considers that these early stages do not develop until about 60% of the ultimate. There is certainly a very marked increase in the amount of structural change above 60% which is reflected by the marked departure from linearity of the stress–strain diagram illustrated in Figure 6.2. The former author maintains that linearity here is an artefact resulting from insensitive testing equipment and that some plastic flow has occurred at levels well below 60% of the ultimate stress.

Compression deformation assumes the form of a small kink in the microfibrillar structure and, because of the presence of crystalline regions in the cell wall, it is possible to observe this feature using polarisation microscopy (Figure 7.16). The sequence of irreversible anatomical changes leading to failure originates in the tracheid or fibre wall at that point where the longitudinal cell is displaced vertically to accommodate the horizontally running ray. As stress and strain increase these kinks become more prominent and increase numerically, generally in a preferred lateral direction, horizontally on the longitudinal–radial plane (Figure 7.17) and at an angle to the vertical axis of from 45° to 60° on the longitudinal–tangential plane. These lines of deformation, generally called a crease and comprising numerous kinks, continue to develop in width and length. At failure, defined in terms of maximum stress, these creases can be observed by eye on the face of the block of timber (Dinwoodie, 1968). At this stage there is considerable buckling of the cell wall and delamination within it, usually between the S_1 and S_2 layers. Models have been produced to simulate buckling behaviour and calculated crease angles for instability agree well with observed angles (Grossman and Wold, 1971).

At a lower order of magnification, Dinwoodie (1974) has shown that the angle at which the kink traverses the cell wall (Figure 7.16) varies systematically

Figure 7.16 Formation of kinks in the cell walls of spruce timber (*Picea abies*) during longitudinal compression stressing. The angle θ lying between the plane of shear and the middle lamella varies systematically between timbers and is influenced by temperature (× 1600, polarised light). (© BRE.)

Figure 7.17 Failure under longitudinal compression at the macroscopic level. On the longitudinal radial plane the crease (shear line) runs horizontally, whereas on the longitudinal tangential plane the crease is inclined at 65° to the vertical axis. (© BRE.)

between earlywood and latewood, between different species, and with temperature. Almost 72% of the variation in the kink angle could be accounted for by a combination of the angle of the microfibrils in the S_2 layer and the ratio of cell wall stiffness in the longitudinal and horizontal planes.

Attempts have been made to relate the size and number of kinks to the amount of elastic strain or the degree of viscous deformation (Dinwoodie and Bonfield, 1993). Under conditions of prolonged loading, total strain and the ratio of creep strain to elastic strain (relative creep), appear to provide the most sensitive guide to the formation of cell wall deformation; the gross creases appear to be associated with strains of 0.33% (Keith, 1972).

The number and distribution of kinks is dependent on temperature and moisture content. Increasing moisture content, even though resulting in a lower strain to failure, results in the production of more kinks, although each is smaller in size than its 'dry' counterpart. These are to be found in a more even distribution than they are in dry timber. Increasing temperature results in a similar wider distribution of the kinks.

7.7.1.3 Static bending

In the bending mode timber is subjected to compression stresses on the upper part of the beam and tensile stresses on the lower part. As the strength of clear timber in compression is only about one third that in tension, failure will occur on the compression side of the beam long before it will do so on the tension side. In knotty timber, however, the compressive strength is often equal to and can actually exceed the tensile strength. As recorded in the previous section, failure in compression is progressive and starts at low levels of stressing. Consequently, the first stages of failure in bending in clear straight-grained timber will frequently be associated with compression failure. Also, as both the bending stress and consequently the degree of compression failure increase, so the neutral axis will move progressively downwards from its original central position in the beam (assuming uniform cross-section), thereby allowing the increased compression load to be carried over a greater cross-section. Fracture occurs when the stress on the tensile surface reaches the ultimate strength in bending.

7.7.1.4 Toughness

Timber is a tough material, and in possessing moderate to high stiffness and strength in addition to its toughness, it is favoured with a unique combination of mechanical properties emulated only by bone which, like timber, is a natural composite.

Toughness is generally defined as the resistance of a material to the propagation of cracks; various methods are available to measure toughness or some index of toughness and these are described in Section 7.2.2.1. In the comparison of materials it is usual to express toughness in terms of *work of fracture*, which is a measure of the energy necessary to propagate a crack, thereby producing new surfaces.

In timber the work of fracture, a measurement of the energy involved in the production of cracks at right angles to the grain, is about 10^4 J/m^2; this value is an order of magnitude less than that for ductile metals, but is comparable with that for the manmade composites. The energy required to break all the chemical bonds in a plane cross-section is of the order of 1–2 J/m^2; that is, four orders of magnitude lower than the experimental values. As pullout of the microfibrils does not appear to happen to any great extent, it is not possible to account for the high work of fracture in this way (Gordon and Jeronimidis, 1974; Jeronimidis, 1980).

One of the earlier theories to account for the high toughness in timber was based on the work of Cook and Gordon (1964) who demonstrated that toughness in fibre-reinforced materials is associated with the arrest of cracks made possible by the presence of numerous weak interfaces. As these interfaces open, so secondary cracks are initiated at right angles to the primary, thereby dissipating its energy. This theory is applicable to timber as Figure 7.18 illustrates,

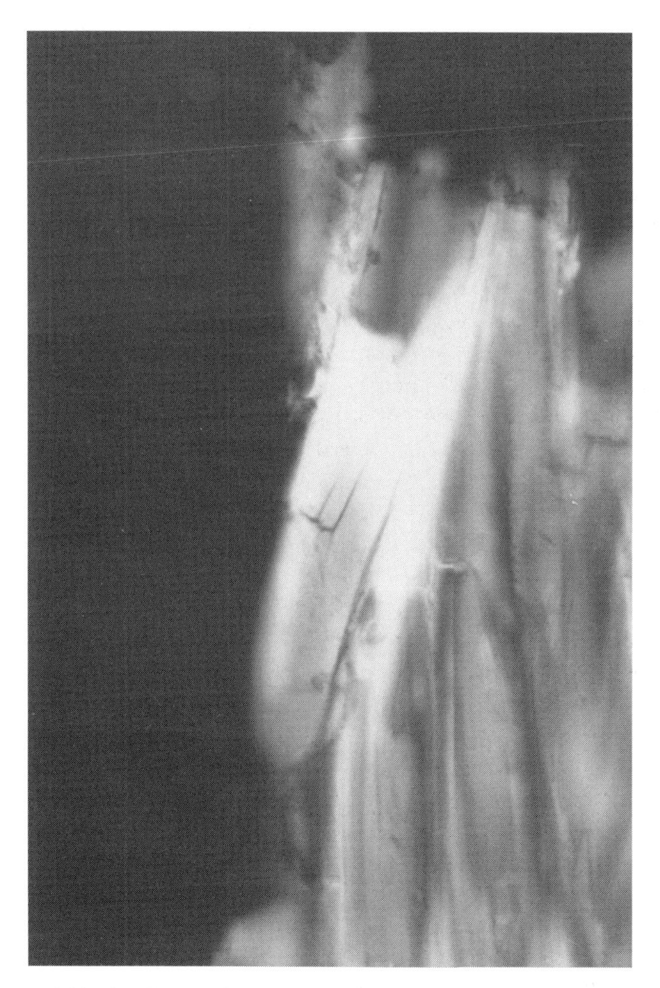

Figure 7.18 Crack-stopping in a fractured rotor blade. The orientation of the secondary cracks corresponds to the microfibrillar orientation of the middle layer of the secondary cell wall (× 990, polarised light.) (© BRE.)

but it is doubtful whether the total discrepancy in energy between experiment and theory can be explained in this way.

Subsequent investigations have contributed to a better understanding of toughness in timber (Gordon and Jeronimidis, 1974; Jeronimidis, 1980). Prior to fracture it would appear that the cells separate in the fracture area; on further stressing these individual and unrestrained cells buckle inwards generally assuming a triangular shape. In this form they are capable of extending up to 20% before final rupture, thereby absorbing a large quantity of energy. Inward buckling of helically wound cells under tensile stresses is possible only because the microfibrils of the S_2 layer are wound in a single direction. Observations and calculations on timber have been supported by glassfibre models and it is considered that the high work of fracture can be accounted for by this unusual mode of failure. It appears that increased toughness is possibly achieved at the expense of some stiffness, as increased stiffness would have resulted from contrawinding of the microfibrils in the S_2.

So far, we have discussed toughness in terms of only clear timber. Should knots or defects be present, timber will no longer be tough and the comments made earlier as to viewpoint are particularly relevant here. The material scientist sees timber as a tough material; whereas the structural engineer will view it as a brittle material because of its inherent defects and this theme is developed in Section 7.7.2.

Loss in toughness, however, can arise not only on account of the presence of defects and knots, but also through the effects of acid, prolonged elevated temperatures, fungal attack, or the presence of compression damage with its associated development of cell wall deformations. These result from overstressing within the living tree, or in the handling or utilisation of timber after conversion (see Section 8.2.4 and Figure 8.1; Dinwoodie, 1971; Wilkins and Ghali, 1987; Koch *et al.*, 1996). Under these abnormal conditions the timber is said to be *brash*.

7.7.1.5 *Fatigue*

Fatigue is usually defined as the progressive damage and failure that occurs when a material is subjected to repeated loads of a magnitude smaller than the static load to failure; it is perhaps the repetition of the loads that is the significant and distinguishing feature of fatigue.

Although there has always been some interest in fatigue in timber, this was mainly academic in the early days; as a fibre composite it was considered that it would not fail in fatigue in the applications in which it was then being used. However, the use of timber for the Mosquito bomber and gliders in the early 1940s, and more recently its use for the manufacture of blades for the new wind-powered generators sharply focused attention on the fatigue properties of timber. These blades are subject not only to cyclic loading from gravity and wind stresses, but also to a complex mixture of loading modes from the natural variability in wind strength and direction.

In fatigue testing the load is applied generally in the form of a sinusoidal or a square wave; minimum and maximum stress levels are usually held constant throughout the test though other wave forms, and block or variable stress levels may be applied. In any fatigue test the three most important criteria in determining the character of the wave form are:

- the stress range, $\Delta\sigma$, where ($\Delta\sigma = \sigma_{max} - \sigma_{min}$).
- the R ratio, where $R = \sigma_{min}/\sigma_{max}$, which is the position of minimum stress (σ_{min}) and maximum stress (σ_{max}) relative to zero stress. This will determine whether or not reversed loading will occur; this is quantified in terms of the R ratio, thus a wave form lying symmetrically about zero load will result in reversed loading and have an R ratio of –1.
- The frequency of loading.

The usual method of presenting fatigue data is by way of the *S–N* curve where log *N* (the number of cycles to failure) is plotted against the mean stress *S*; a linear regression is usually fitted (Figure 7.19).

Using test pieces of Sitka spruce, laminated Khaya and compressed beech, Tsai and Ansell (1990) carried out fatigue tests under load control in four-point bending. The tests were conducted in repeated and reversed loading over a range of five *R* ratios at three moisture contents. Fatigue life was found to be

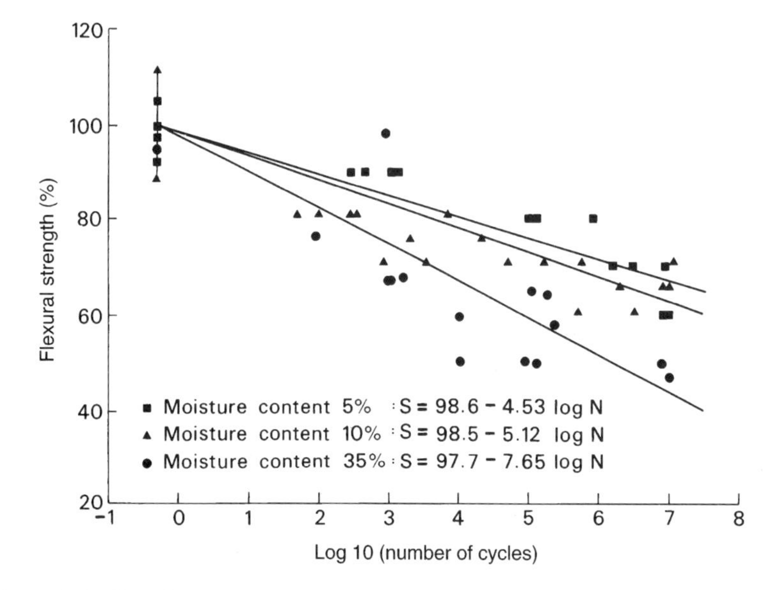

Figure 7.19 The effect of moisture content on sliced Khaya laminates fatigued at R = 0. The maximum peak at stresses are expressed as a percentage of static flexural (bending) strength. (From Tsai, K.T. and Ansell, M.P. (1990) *J. Mat. Sci.*, **25**, 865–878, reproduced by permission of Kluwer Academic Publishers.)

largely species independent when normalised by static strength, but was reduced with increasing moisture content (Figure 7.19) and under reversed loading. The accumulation of fatigue damage was followed microscopically in test pieces fatigued at $R = 0.1$ and was found to be associated with the formation of kinks in the cell walls and compression creases in the wood (see Section 7.7.1.2).

In related work Bonfield and Ansell (1991) investigated the axial fatigue in constant amplitude tests in tension, compression and shear in both Khaya and Douglas fir (Figure 7.20) using a wide range of R ratios and confirmed that reversed loading is the most severe loading regime. Fatigue lives measured in all-tensile tests ($R = 0.1$) were considerably longer than those in all-compression tests ($R = 10$), a result which they related to the lower static strength in compression relative to tension. *S–N* data at different R ratios yielded a set of constant life lines when alternating stress was plotted against mean stress; these lifelines possessed a point of inflection when loading became all compressive.

MODELLING AND FAILURE CRITERIA IN FATIGUE

By far the simplest and most frequently used expression of fatigue data is the use of linear regressions to describe *S–N* relationships. (Some examples are:

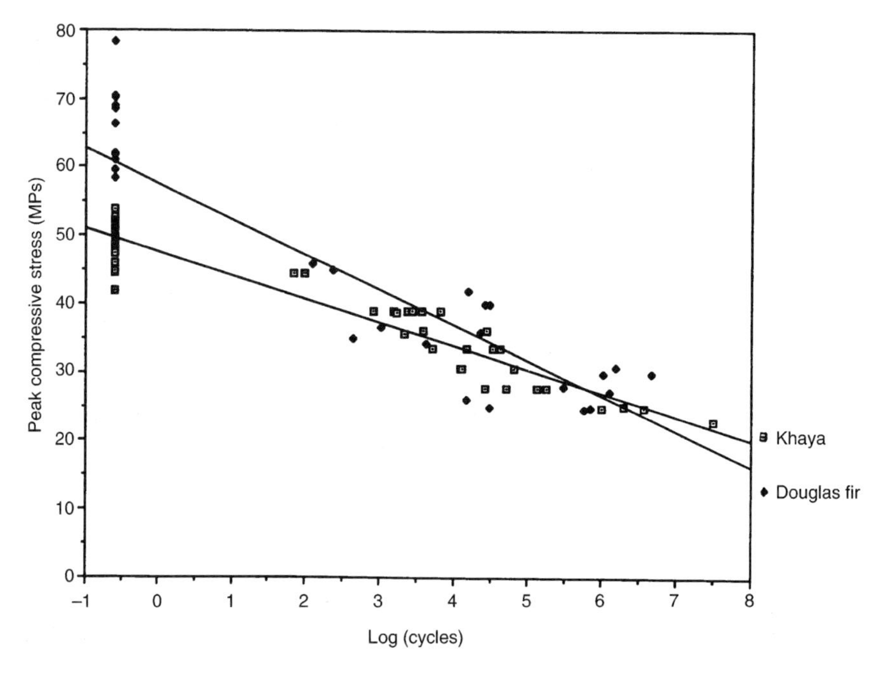

Figure 7.20 *S–N* curves for Khaya and Douglas fir. The data points at $\log_{10} N = -0.6$ (one quarter of one cycle) are the static values. (From Bonfield, P.W. and Ansell, M.P. (1991) *J. Mat. Sci.*, **26**, 4765–4773, reproduced by permission of Kluwer Academic Publishers.)

McNatt (1978) for particleboard in tension and shear; Tsai and Ansell (1990) for Sitka spruce, Khaya and beech in bending; Bonfield and Ansell (1991) for Khaya and Douglas fir in tension, compression and shear; and Thompson *et al.* 1996 for particleboard in bending. It is interesting to note that the regressions of McNatt and Thompson are remarkably similar.)

Mohr (1996), however, in evaluating fatigue data from the literature as well as from international data bases, found that a double linear regression line fitted to the S–$\log N$ data gave a better fit than the conventional linear plot. Additionally, for most loading modes, a log–log plot fitted the data even better than both the semilogarithmic plots. Mohr also proposed a fatigue factor

$$k_{fat} = f_{kn}/f_k \tag{7.17}$$

where f_k is the five percentile short-term (static) strength and f_{kn} is the five percentile fatigue strength for N cycles. When $N = 10^7$ cycles and the R ratio is –1 (i.e. reversed loading), k_{fat} in tests in bending was of the order of 0.30.

In a second approach to modelling, this time on data from spruce subjected to cyclic loading at 80% of its short-term strength using a square wave form at four different frequencies, Clorius *et al.* (1996) found that the number of cycles to failure was a poor measure of fatigue performance. An initial model based on *total work* during the fatigue life was rejected, to be replaced by one that better fitted the experimental data; this second model was based on the elastic, viscous and non-recovered viscoelastic *work*. It is interesting to note that the accumulated creep was identified with the formation of kinks in the cell wall (see Section 6.3.1.6).

A third approach to modelling has been the application of the statistical theory of the absolute reaction rate. Using existing data from constant load tests, Liu *et al.* (1994) derived the parameters in their new mathematical model to describe the fatigue life of structural timber under sinusoidal loading. Their model predicts that fatigue life under a cyclic bending load will be independent of stress frequency: under isothermal conditions, fatigue life depends only on the mean stress and a function of the stress amplitude of the applied load.

The fourth type of modelling of fatigue is the use by Nielsen (1997) of his DVM (damage viscoelastic material) theory which was originally developed to predict time to failure under static load (see Section 7.7.2.1).

7.7.2 Engineering approach to strength and fracture

Two basic aspects will be discussed: the first is *fracture mechanics*, which is concerned with the strength of actual materials as limited by the propagation of cracks; the second, in assessing variability in materials, relates strength to the weakest link in the structure of the material. The two concepts are obviously related though each may be applied independently to different aspects of strength.

7.7.2.1 Fracture mechanics and its application to timber

Fracture mechanics is the applied mechanics of crack growth starting with a flaw or crack. The concept of crack growth is attributed to Griffith who postulated, first, that in a material there are flaws and it is the worst flaw which causes failure and, second, that a balance exists between the energy due to elastic strain and the energy necessary to produce new surfaces. The largest flaw will become self-propagating when the rate of release of strain energy exceeds the rate of increase of surface energy of the propagating crack. The theory was later modified to take account of both plastic work and kinetic energy.

Now assuming a homogeneous, isotropic, continuum structure and that all deformation around the crack tip obeys linear elasticity theory, it follows from Griffith's concept that the crack will propagate when sufficient stress is available to operate a suitable mechanism of fracture, and the strain released by an increment of crack growth must equal or exceed the energy required to form the new crack surfaces. This occurs when

$$\sigma^2 \pi c = EG_c \tag{7.18}$$

where $2c$ is the crack length, σ is the stress applied, E is the modulus of elasticity, and G_c is the critical strain energy release rate ($G_c = 2\gamma$, where γ is the energy required to form new fracture surfaces). The left-hand side of equation (7.18) contains engineering parameters (stress + distance) while the right-hand side contains material properties. As

$$EG_c = K_c^2 \tag{7.19}$$

for plane stress (or $EG_c/(1-\nu^2)$ for plane strain, where ν is the Poisson's ratio), it follows that K_c, the *stress intensity factor* at the critical state at which the crack will grow, is also a material property. Most importantly, K_c is independent of crack length and this means for each material it is possible to obtain the relationship between the critical stress and the length of crack.

By measuring stress to failure for a series of crack geometries, it is possible by plotting $\sigma \times 1/c$ to obtain K_c experimentally; alternatively, assuming complete confidence in the equation, a test at a single crack length will suffice. Knowing K_c and setting limits on c, the limiting stress σ can be calculated from equations (7.18) and (7.19).

Three modes of failure are defined: the opening mode is usually denoted by I and the appropriate critical intensity factor as K_{IC}, the forward shear mode is II and the transverse shear mode is III.

For an infinite plate with a crack length of $2c$,

$$K_c = \sigma (\pi c)^{1/2} \tag{7.20}$$

Although timber can be treated as an elastic material at low levels of applied stress for short periods of time (as explained in Chapter 6), it is surprising to find that the application of linear elastic fracture mechanics to timber did not take place until as late as 1961, when it was shown that the product of fracture stress and the square root of crack length was approximately constant. Subsequent workers confirmed the independence of the stress intensity factor (K) from crack length, thereby indicating that 'fracture toughness' (the critical stress intensity factor, K_c) could be as useful as a wood constant as it has become for other materials.

However, unlike isotropic materials where there are only three fracture toughness values (K_{Ic}, K_{IIc}, and K_{IIIc}) corresponding to the three principal modes of crack propagation, timber is modelled as an orthotropic material and requires six values of fracture toughness for each mode, thereby making a total of 18. Due to the complexity of the situation much of the work on fracture toughness of timber has been done using the isotropic solution, a situation which under certain conditions, as when the crack coincides with one of the three principal planes, gives satisfactory results. One of the major problems associated with orthotropic materials is that the crack does not necessarily propagate in a direction perpendicular to the direction of maximum stress. Thus in timber, macroscopic crack extension almost always occurs parallel to the grain even though it is initiated in a different plane, thereby giving rise to a mixed mode type of failure.

The value of K_c (fracture toughness) is dependent not only on orientation as inferred above, but also on the opening mode, wood density, moisture content, specimen thickness, and crack speed. There is some evidence to show that it may also vary with crack length.

EFFECT OF ORIENTATION

The value of K_{Ic} (K_c in mode I) in the four weak parallel-to-the-grain systems is about one-tenth that in the two tough across-the-grain systems; thus Schniewind and Centeno (1973) found values for K_{Ic} of 309–410 $kNm^{-3/2}$ for the four longitudinal systems, and 2417 and 2692 $kNm^{-3/2}$ for the two transverse systems in Douglas fir timber.

OPENING MODE

Fracture of timber can occur as frequently in mode II as in mode I. Two very different types of test pieces have been used in the determination of K_{IIc}, these being a single edge-notched beam (Leicester, 1974; Walsh, 1972), and an end-notched beam (Barrett and Foschi, 1977a; Fonselius, 1991). Values of K_{IIc} for propagation in the longitudinal–radial plane for four softwoods range from 1626 $kNm^{-3/2}$ for amaphilis fir to 2187 $kNm^{-3/2}$ for lodgepole pine (Barrett, 1981).

No information on mode III is available.

EFFECT OF DENSITY

The little information that is available indicates that K_c increases with increasing density.

MOISTURE CONTENT

The value of K_c decreases with an increase in moisture content (Ewing and Williams, 1979) or under changing moisture content (Fonselius, 1991).

SPECIMEN THICKNESS (SIZE EFFECT)

The value of K_c is affected by specimen thickness; a number of studies have indicated that thicker specimens have lower values of fracture toughness (Ewing and Williams 1979).

Aicher (1992), following tests on spruce notched in the longitudinal–tangential plane and loaded in bending, applied both linear and non-linear fracture mechanics to his results and concluded that, whereas linear fracture mechanics is applicable to larger-dimensioned timber, non-linear fracture mechanics must be applied to small-dimensioned timber in which the fracture process has become more significant.

CRACK SPEED

The value of K_c increases with increasing crack speed (Ewing and Williams, 1980).

Fracture mechanics has been applied to various aspects of wood behaviour and failure, of which the following are selected examples:

- The effect of knot size on the tensile strength of timber has been established by the application of fracture mechanics. By treating the knots as cracks, it was possible to predict ultimate strength; the values obtained agreed well with data from actual tensile tests (Pearson, 1974).
- The effect of drying splits on the tensile strength perpendicular to the grain of timber was determined using fracture mechanics and treating the splits as cracks; very good agreement was obtained between actual and predicted values (Schniewind and Lyon, 1973).
- The effect of the presence of butt joints in laminated beams has also been accurately predicted using fracture mechanics, by treating the joint as a crack (Smith and Penney, 1980).
- End splits which can arise during the drying of timber can have a marked influence in determining the shear capacity of a beam. A mode II fracture criterion has been developed to predict the effect of end splits on shear capacity (Barrett and Foschi 1977b).

- The effect of ray spacing in determining the fracture toughness of timber has been investigated in tension parallel to the grain in un-notched aspen test pieces in mode I, because observations in the LR plane of fracture had shown the ray–fibre interface to be a notable zone of crack propagation. Using a model based on inter-ray distance, the calculated K_{Ic}^{LR} was close to published data (Akande and Kyanka, 1990).
- The realisation in the 1970s and 1980s that the method of determining duration of load from small clear test pieces was inappropriate for structural timber (described in Section 7.6.12.2) led to the application of fracture mechanics to determine duration of load. Interest focussed on *slow* crack growth as it was argued that the long-term strength of wood (i.e. the duration of load effect described earlier) is dependent on the growth of subcritical cracks which are too small to cause failure initially (Nadeau *et al.*, 1982). Time-dependent crack growth has been studied using viscoelastic fracture mechanics (Mindness *et al.*, 1975, 1980; Johns and Madsen, 1982; McDowell, 1984; Nielsen 1986). Prediction of time to failure (DOL) from such a model agreed well with experimental results of duration of load tests on Douglas fir specimens (McDowell, 1984). Johns and Madsen (1982) applied a viscoelastic limited-ductility fracture mechanics model which had been developed from an earlier model by Nielsen; the model has been shown to fit experimental results for three independent series of tests. The model was refined by Nielsen (1986, 1997) to become his *damaged viscoelastic material* (DVD) theory. This theory, which is based on the premise that timber is a cracked viscoelastic material, is made general with respect to type of defects and is applicable to both DOL and fatigue modelling.
- Fracture mechanics has also been used to study the effects of rate of loading (Nadeau *et al.*, 1982), and long-term failure under fatigue (Nielsen, 1997; see previous point above).

7.7.2.2 *Weibull distribution*

The Weibull theory is a statistical approach to the failure of a brittle material. Using the weak-link concept, Weibull presented the first theories capable of quantifying effects of stress distribution and volume on strength of materials. The distribution which arises from a set of non-linear hazard functions is called the Weibull distribution. This is an extreme-value distribution for minimum values and consequently is strongly influenced by the lowest value unlike most other distributions. It is to be preferred for predicting the lower end of a distribution which quite frequently is the critical area in practical terms.

Two and sometimes three parameters are required covering the shape, scale and location of the distribution; although methods are established for estimating these parameters, no one method appears ideal under all circumstances. Computer programs for calculating these parameters have been published (Pierce, 1976).

The complexity of the concept has done little to prevent its fairly extensive usage in ceramics, but so far it has been used only infrequently on timber.

One of the first specialised applications of the Weibull distribution was in the evaluation of the bending strengths of beams where it is found that the changes in strength resulting from changes in span and beam depth could be explained satisfactorily by a weakest-link model. It has also been successfully applied to predict the relation between specimen volume and strength in timber stressed in tension perpendicular to the grain (Barrett, 1974). Timber appears to exhibit a size effect to a greater extent than that in most other materials. A less specialised application has been its use on one or two occasions for the derivation of working stresses.

7.7.2.3 Duration of load

The philosophy of duration of load (DOL) changed in the late 1970s when it was realised that the measurement of duration of load on small, clear test pieces gave unrealistic predictions for structural-size timber.

The application of damage-accumulation models and linear viscoelastic fracture mechanics are now well-established tools for the prediction of DOL. These techniques have been discussed in some detail in Sections 7.6.12.2 and 7.7.2.1, respectively.

7.7.2.4 Fatigue

The classical approach to fatigue has now been complemented by an engineering approach in the form of linear elastic fracture mechanics; this is discussed in Sections 7.7.1.5 and 7.7.2.1.

7.8 Practical significance of strength data

The engineer contemplating the use of timber will automatically ask:

- how do the mechanical properties of timber compare with other materials?
- are there working stresses for timber and, if so, how are they derived and how can they be used?

These questions are discussed below.

7.8.1 Comparison of strength data of timber and other materials

Timber has been used as a construction material for centuries and continues to be used extensively in many forms of construction despite the increase in the number of competing materials. Its ability to withstand competition is a reflection not just of the level of strength it possesses, but more importantly the level of strength in terms of both cost and weight.

In Table 7.6 the tensile and compression strengths of whitewood, one of the principal timbers of construction, are set out in comparison with many other traditional materials together with some of the new constructional materials; stiffness (in bending) is included for completeness. The timber values were derived using small clear test pieces. It will be observed that the tensile strength of whitewood is in the middle of the range, but when strength is quoted in terms of specific gravity so that the strength of an equal mass of material is compared, timber is far superior to the traditional materials, though not as good as the manmade composites. In both stiffness, which at best is only moderate in comparison to other materials, and also compression parallel to the grain, which is poor in comparison, the position of timber is enhanced when the property is expressed in terms of its specific gravity (SG). The use of timber for the Mosquito aircraft during the war and its continued, though declining, use for glider production bears testament to its high strength to mass ratio.

When cost is taken into the comparison process, the position of timber relative to other materials is certainly favourable. In the comparison of timber and steel beams to carry the same load with similar deflections the cost of the former will be slightly lower than its steel counterpart. However, where strength and stiffness are not so important and where timber has to be machined to a particular profile and subsequently painted, the competition from extruded rigid plastic foams is very acute and, as labour costs escalate, timber will continue to be replaced.

Table 7.6 Strength and stiffness of timber (whitewood small clears) in comparison with other materials

Material	*Specific gravity*	*Tensile strength (MN/m^2)*	*Specific tensile strength (MN/m^2)*	*Modulus of elasticity (GN/m^2)*	*Specific modulus (GN/m^2)*	*Compression strength (MN/m^2)*
Timber	0.46	104	226	10	22	37
Concrete	2.5	4	2	48	19	69
Glass	2.5	50	20	69	28	50
Duralumin	2.8	247	88	69	25	–
Cast iron	7.8	138	18	207	26	120
Steel (mild; 0.06% C)	7.9	459	58	203	26	800
Acrylonitrite–butadiene –styrene (ABS)	1.1	50	45	3	3	50
Rigid poly(vinyl chloride)	1.5	59	39	2.4	1.7	55
Polyester resin/G cloth	1.8	276	153	18	10	270
Epoxy uni-directional roving[a]	1.8	1100	611	45	25	400
Carbon fibre reinforced plastic (CFRP)	1.5	1040	693	180	120	1040

[a] A composite comprising an epoxide resin in which are embedded bundles of glass strands lying parallel to each other

7.8.2 Design stresses for timber

7.8.2.1 Background

Timber, like many other materials, is graded according to its anticipated performance in service; because of its inherent variability, distinct grades of material must be recognised.

GRADE STRESSES IN UK PRIOR TO 1973

These were derived from the testing of small clear test pieces (Section 7.2.2.1) as set out in Table 7.7 and described fully in the 1981 version of this text, as well as in Desch and Dinwoodie (1996).

GRADE STRESSES IN UK FROM 1973 TO 1995

The realisation in the 1970s that the duration of load values derived from the testing of small clear test pieces were not appropriate for structural timber led to the derivation of grade stresses directly from actual structural-size timber.

This approach necessitated the introduction of grading of the timber by either visual or mechanical means about which more is said below. In order to define these grade classes in terms of actual strength, tests on structural timber had to be performed which led to the derivation of actual grade stresses either in the form of strength classes, or as grade stresses for individual species and grades (see Table 7.7 and Desch and Dinwoodie, 1996). The advantage of the strength class system over the listing of stresses for individual species and grades is that it allows suppliers to meet a structural timber specification by supplying any combination of species and grade listed in BS 5268 Part 2 (1996) as satisfying the specified class.

Structural design using these stresses is by way of *permissible stress* design according to BS 5268 Part 2 (1996).

7.8.2.2 Characteristic values for structural timber in Europe (including UK) from 1996

The formation of the EC as a free trade area and the production of the *Construction Products Directive* (CPD) led automatically to the introduction and implementation of new European standards and the withdrawal of conflicting National standards. Such an approach has much merit, but as far as the UK is concerned, it has led to changes in both the derivation and use of working stresses. First, test results are now expressed in terms of a *characteristic value* expressed in terms of the lower 5th percentile, in contrast to the former use in the UK of a mean value and its standard deviation. Second, in the design of timber structures the new Eurocode 5 (see below) is written in terms of *limit*

Table 7.7 Changes in derivation of design stresses over the period before 1973 to about 2005

UK			EUROPE
→1973	**1973–1995**	**1995 – about 2005**	**1996 →**
Data from testing small clear test pieces	Structural timber	Structural timber	Structural timber
↓	↓	↓	↓
Mean value less 2.33s	Grading Tests°	Grading Tests°	Grading Tests°
↓	↓ ↓ BS 5820	↓ ↓ EN 408	↓ ↓ EN 408
Divided by safety factor* to give **Basic stress**	Visual BS 4978 (1988) BS 5756 (1980) Machine BS 4978 (1988) ↓ ↓ ↓ ↓ ↓	Visual BS 4978 (1996) BS 5756 (1997) Machine EN 519 EN 384 ↓ ↓ ↓ ↓	Visual EN 518 Machine EN 519 EN 384 ↓ ↓ ↓ ↓
↓	↓ ↓ ↓	↓ ↓ ↓	↓ ↓ ↓
Multiplied by *strength ratio*	**Grade** or **Strength classes** ↓ BS 4978 (1988) ↓	↓→ **Strength classes** and ↓ **Characteristic values** ↓ EN 338	**Strength classes** **and** **Characteristic Values** EN 338, EN 1912 ↓
↓	↓ ↓		
Derived **Grade stresses** in **BS 5268 Part 2**	Stresses in standard **BS 5268 Pt. 2**	↓ ↓ ↓ factored ↓ ↓ → **Grade stresses**	
↓	↓	↓	↓
Design code BS 5268 Pt. 2	**Design code** BS 5268 Pt. 2	**Design code** BS 5268 Pt. 2	**Design code** ENV 1995–1–1

s = standard deviation
* = safety factor (1.4 for compression ∥ to the grain; 2.25 for all other modes) to cover effects of specimen size and shape, rate of loading and duration of load
° = testing used only to derive values for inclusion in the standards

state design, in contrast to the former UK use of permissible stress design. Third, the number of strength classes in the new European system is greater than in the UK system, thereby giving rise to mismatching of certain timbers. An example of the new European strength class system is given in Table 7.8.

However, apart from these three major changes, the grading of timber into strength classes and the original characterisation of these by testing adopts a similar procedure to that used in the UK after 1973 (see Table 7.7).

Characteristic values for structural timber in a range of strength classes is to be found in EN 338, while the corresponding values for the load-bearing wood-based panels are located in EN 12369 with the exception of plywood: because of the great diversity of plywood types many of which are defined by National boundaries, characteristic values for plywood should be taken from manufacturers' literature, provided there is reference to first, that these values have been obtained by testing to EN 789 (1996) and EN 1058 (1996), and second, that production was in control and verified by a third party.

As noted above, the design of timber structures must be in accordance with Eurocode 5. The characteristic values for both timber and wood-based panel products must be reduced according to the period of loading and service class

Table 7.8 An extract from Table 1 of EN 338 illustrating for each of the strength and stiffness parameters the characteristic values for certain selected strength classes

Parameter		*Strength class and characteristic values*					
		Softwood				*Hardwood*	
		C16	*C18*	*C22*	*C24*	*D30*	*D40*
Strength properties (in N/mm²)							
Bending	$f_{m,k}$	16	18	22	24	30	40
Tension parallel	$f_{t,0,k}$	10	11	13	14	18	24
Tension perpendicular	$f_{t,90,k}$	0.3	0.3	0.3	0.4	0.6	0.6
Compression parallel	$f_{c,0,k}$	17	18	20	21	23	26
Compression perpendicular	$f_{c,90,k}$	4.6	4.8	5.1	5.3	8.0	8.8
Shear	$f_{v,k}$	1.8	2.0	2.4	2.5	3.0	3.8
Stiffness properties (in kN/mm²)							
Mean modulus of elasticity parallel	$E_{0,mean}$	8	9	10	11	10	11
5% modulus of elasticity parallel	$E_{0,05}$	5.4	6.0	6.7	7.4	8.0	9.4
Mean modulus of elasticity perpendicular	$E_{90,\ mean}$	0.27	0.30	0.33	0.37	0.64	0.75
Mean shear modulus	G_{mean}	0.50	0.56	0.63	0.69	0.60	0.70
Density (in kg/m³)							
Density	ρ_k	310	320	340	350	530	590
Average density	ρ_{mean}	370	380	410	420	640	700

(defined in terms of level of relative humidity). Values of K_{mod} (duration of load and service class) and K_{def} (creep and service class) are set out in Eurocode 5 for each of the three service classes.

7.8.2.3 UK position from 1996 to about 2005

This is the transition period for all European countries in which a choice is available between their existing national grading and design standards and the new European system. In theory both systems exist side by side, but in practice many countries, including the UK, have wisely taken advantage of this period to modify their existing standards in order to bring them nearer the European system and reduce the magnitude of the change at the termination of this period.

The UK has already adopted machine grading of softwoods to EN 519 (1995) and has redrafted BS 4978 (1996) so that it now relates only to the visual grading of softwoods. Machine graded timber is assigned to strength classes contained in EN 338 (1995). From this standard, the characteristic value for a particular strength class and material property are read off and in turn converted (factored) to a grade stress for use in the continuing permissible stress design system set out in BS 5268 Part 2 (1996). Visually graded timber is still assigned to grade stresses directly (see Table 7.7).

7.8.2.4 Visual grading

Visual grading as the title implies is a visual assessment of the quality of a piece of structural timber. This is carried out against the permissible defects limits given in standards BS 4978 (1996) and BS 5756 (1997) which conform to EN 518 (1995). However, visual grading is a laborious process because all four faces of the timber should be examined. Furthermore, it does not separate naturally weak from naturally strong timber and hence it has to be assumed that pieces of the same size and species containing identical defects have the same strength: such an assumption is invalid and leads to a most conservative estimate of strength.

7.8.2.5 Machine grading

Many of the disadvantages of visual grading can be removed by machine grading, a process which was introduced commercially in the early 1970s. The principal underlying the process is the high correlation that has been found to exist between the moduli of elasticity and rupture; this was described in Section 7.5.1 and illustrated in Figure 7.5. Grading machines based on this relationship usually provide higher yields of the higher grades than are achieved by visual grading. As the above relationship varies among the different species, it is necessary to set the grading machine for each species or species group.

The equation of the regression line is only one of a number of inputs to the mathematical model used to determine the machine settings for a particular timber. These settings also depend on:

- bandwidth (separation of the grade boundaries on the x-axis) which in turn is related to the grade combination being graded;
- the overall mean and standard deviation of the modulus of elasticity of the species;
- the interaction of bandwidth and the MOE parameters;
- the cross-sectional size, and whether the timber piece is sawn or planed all round (PAR).

The mathematical model can be designed to select a number of grades or strength classes.

Having established the basic relationship for each species, the grading machine can then be set up to grade the timber automatically. Depending on the type of machine, as each length of timber passes through, it is either placed under a constant load and deflection is measured, or subjected to a constant deflection and load is measured. A small computer quickly assesses the appropriate stiffness-indicating parameter every 100–150 mm along the length of the timber. The lowest reading of load, or the highest reading of deflection, is compared with the preset values stored in the computer and this determines the overall grade of the piece of timber. Grade stamps are then printed on one face of the timber towards one end of the piece.

Under the new European system this grade mark will include the specification number used in the grading (EN 519 (1995))

- the species or species group
- the timber condition (green or dry)
- the strength class
- the grader or company name
- the company registration number
- the certification body (logo).

New grading systems have recently been introduced, one of which, the dynograder scanning system, has been shown to increase both throughput and yield.

References

Standards and specifications

ASTM Standard D143–52 (1972) *Standard methods of testing small clear specimens of timber*, American Society for Testing Materials.

BS 373 (1986) *Methods of testing small clear specimens of timber*, BSI, London

BS 4978 (1988) *Softwood grades for structural use*, BSI, London.
BS 4978 (1996) *Visual grading of softwood*, BSI, London.
BS 5268 *Structural use of timber* Part 2 (1996) – Code of practice for permissible stress design, materials and workmanship, BSI, London.
BS 5756 (1997) *Visual strength grading of hardwoods*, BSI, London.
BS 5820 (1979) *Methods of test for determination of certain physical and mechanical properties of timber in structural sizes*, BSI, London.
EN 338 (1995) *Structural timber – Strength classes.*
EN 384 (1995) *Structural timber – Determination of characteristic values on mechanical properties and density.*
EN 408 (1995) Timber structures – Structural timber and glued laminated timber – Determination of some physical and mechanical properties.
EN 518 (1995) *Structural timber – Grading – Requirements for visual strength grading standards.*
EN 519 (1995) *Structural timber – Grading – Requirements for machine strength graded timber and grading machines.*
EN 789 (1996) *Timber structures – Test methods – Determination of mechanical properties of wood-based panels.*
EN 1058 (1996) *Wood-based panels – Determination of characteristic values of mechanical properties and density.*
ENV 1156 (1999) *Wood-based panels – Determination of duration of load and creep factors.*
EN 1912 (1998) *Structural timber – Strength classes – Assignment of visual grades and species.*
EN 12369 (not yet issued) *Wood-based panels – Characteristic values for use in structural design.*
ENV 1995–1–1 (1994) *Eurocode 5: Design of timber structures. Part 1.1. General rules and rules for buildings.*

Literature

Anon (1996) Specifying structural timber, *Digest 416, Building Research Establishment, Watford*, 8pp.
Aicher, S. (1992) Size effects in linear and non-linear fracture mechanics, in *Proc. Workshop on Fracture Mechanics in Wooden Materials, Bordeaux*, Ed. P. Morlier, G. Valentin and I. Seoane, published by the management committee of EC COST 508 Action, 195–203.
Akande, J.A. and Kyanka, G.H. (1990) Evaluation of tensile fracture in aspen using fractographic and theoretical methods, *Wood and Fiber Sci.*, **22**(3), 283–297.
Barrett, J.D. (1974) Effect of size on tension perpendicular to grain strength of Douglas fir, *Wood and Fiber*, **6** (2), 126–143.
Barrett, J.D. (1981) Fracture mechanics and the design of wood structures, *Phil. Trans. Roy. Soc., Lond.*, **A299**, 217–226.
Barrett J.D. (1996) Duration of load – the past, present and future, in *Proc. of International Conference on Wood Mechanics,* Stuttgart, Germany, Ed. S Aicher, and published by the management committee of EC COST 508 Action, 121–137.
Barrett, J.D. and Foschi, R.O. (1977a) Mode II stress intensity factors for cracked wooden beams, *Eng. Fract. Mech.*, **9**, 371–378.

Barrett, J.D. and Foschi, R.O. (1977b) Shear strength of uniformly loaded dimension lumber. *Can. J. Civil Eng.*, **4**, 86–95.

Barrett, J.D. and Foschi, R.O. (1978) Duration of load and probability of failure in wood, Part 1: Modelling creep rupture. *Can. J. Civil Eng.*, **5** (4), 505–514.

Barrett, J.D. and Lau, W. (1991) Bending strength adjustment for moisture content for structural lumber. *Wood Sci. Technol.*, **25**, 433–447.

Baumann, R. (1922) Die bisherigen Ergebnisse der Holzprufungen in der Materialprufungsanstalt an der Tech Hochschule Stuttgart. *Forsch. Gebiete. Ingenieurw.*, H231, Berlin.

Bonfield, P.W. and Ansell, M.P. (1991) Fatigue properties of wood in tension, compression and shear. *J. Mat. Sci.*, **26**, 4765–4773.

Burger, N. and Glos, P. (1996) Size effect on tensile strength of timber. *Holz als Roh- und Werkstoff*, **54**(5), 333–340

Caulfield, D.F. (1985) A chemical kinetics approach to the duration of load problem in wood. *Wood & Fiber Sci.*, **17**(4), 504–521.

Cave, I.D. (1969) The longitudinal Young's modulus of *Pinus radiata*. *Wood Sci. Technol.*, **3**, 40–48.

Clorius, C.O., Pedersen, M.U., Hoffmeyer, P. and Dankilde, L. (1996) Fatigue damage of wood, in *Proc. International Conference on Wood Mechanics,* Stuttgart, Germany, Ed. S.Aicher, and published by the Management Committee of EC COST 508 Action, 227–242.

Clouser, W.S. (1959) *Creep of small wood beams under constant bending load*, Report 2150, FPL Madison.

Cook, J. and Gordon, J.E. (1964) A mechanism for the control of crack propagation in all brittle systems, *Proc. Roy. Soc. London*, **A 282**, 508.

Desch, H.E. and Dinwoodie, J.M. (1996) *Timber – structure, properties, conversion and use,* 7th edn, Macmillan, Basingstoke.

Dinwoodie, J.M. (1968) Failure in timber, Part I: Microscopic changes in cell wall structure associated with compression failure. *J. Inst. Wood Sci.*, **21**, 37–53.

Dinwoodie, J.M. (1971) Brashness in timber and its significance. *J. Inst. Wood Sci.*, **28** 3–11.

Dinwoodie, J.M. (1974) Failure in timber, Part II: The angle of shear through the cell wall during longitudinal compression stressing. *Wood Sci. Technol.*, **8**, 56–67.

Dinwoodie, J.M. and Bonfield, P.W.(1993) Plasticity in compression in terms of kink formation, in *Proc. of Workshop on Wood: Plasticity and Damage*, Limerick, Ireland, Ed. G. Birkinshaw, P. Morlier, and I. Seoane, and published by the Management Committee EC COST 508 Action, 101–109.

Easterling, K.E., Harrysson, R., Gibson, L.J. and Ashby, M.F. (1982) The structure and mechanics of balsa wood. *Proc. Roy. Soc., London*, **383**, 31–41.

Ewing, P.D. and Williams, J.G. (1979) Thickness and moisture content effect in fracture toughness of Scots pine. *J. Mat. Sci.*, **14**, 2959–2966.

Ewing, P.D. and Williams, J.G. (1980) Slow crack growth in softwoods, *Proc. Third International Conference on Mechanical Behaviour of Materials,* Cambridge, 1979, Ed. K J. Miller and R.F. Smith, Vol. 3, pp. 293–298.

Fonselius, M. (1991) *Long-term fracture toughness of wood*, Research Report 718, Technical Research Centre of Finland.

Foschi, R.O. and Yao, F.Z. (1986a) Another look at the three duration of load models, in *Proc. XVII IUFRO Congress*, Florence, Italy, paper 19–9–1.

Foschi, R.O. and Yao, F.Z. (1986b) Duration of load effects and reliability based design, in *Proc. XVII IUFRO Congress*, Florence, Italy, paper 19–1–1.

Fridley, K.J., Tang, R.C. and Soltis, L.A. (1989) Thermal effects on load-duration behaviour of lumber. Part I: Effect of constant temperature. *Wood & Fiber Sci.*, **21** (4), 420–431

Fridley, K.J, Tang, R.C. and Soltis, L.A. (1991) Moisture effects on the load-duration behaviour of lumber. Part I: Effect of constant relative humidity. *Wood & Fiber Sci.*, **23** (1), 114–127.

Fridley, K.C., Tang, R.C. and Soltis, L.A. (1992a) Load-duration effects in structural lumber: strain energy approach. *J. Struct. Eng., Structural Div. ASCE*, **118** (9), 2351–2369.

Fridley, K.J, Tang, R.C. and Soltis, L.A. (1992b) Moisture effects on the load-duration behaviour of lumber. Part II: Effect of cyclic relative humidity. *Wood & Fiber Sci.*, **24** (1), 89–98.

Fridley, K.C., Tang, R.C., Soltis, L.A. and Yoo C.H. (1992c) Hygrothermal effects on load-duration behaviour of structural lumber. *J. Struct. Eng., Structural Div. ASCE*, **118** (4), 1023–1038.

Gerhards, C.C. (1979) Time-related effects on wood strength: A linear-cumulative damage theory. *Wood Sci.*, **11** (3), 139–144.

Gerhards, C.C. (1982) Effect of moisture content and temperature on the mechanical properties of wood: an analysis of immediate effects. *Wood and Fiber*, **14** (1), 4–26.

Gerhards, C.C. and Link C.L. (1987) A cumulative damage model to predict load duration characteristics of lumber. *Wood and Fiber Sci.*, **19** (2), 147–164.

Gordon, J.E. and Jeronimidis, G. (1974) Work of fracture of natural cellulose. *Nature* (London), **252**, 116.

Green, D.W., Link, C.L., De Bonis, A.L. and McLain, T.E. (1986) Predicting the effect of moisture content on the flexural properties of Southern pine dimension lumber. *Wood Fiber Sci.*, **18**, 134–156.

Grossman, P.U.A. and Wold, M.B. (1971) Compression fracture of wood parallel to the grain, *Wood Sci. Technol.*, **5**, 147–156.

Hoffmeyer, P (1990) *Failure of wood as influenced by moisture and duration of load.* State University of New York, Syracuse, New York. USA.

James, W.L. (1962) Dynamic strength of elastic properties of wood. *For. Prod. J.*, **11** (9), 383–390.

Jeronimidis, G. (1980) The fracture behaviour of wood and the relations between toughness and morphology. *Proc. Roy. Soc., London*, **B208**, 447–460.

Johns, K. and Madsen, B. (1982) Duration of load effects in lumber. Part 1: A fracture mechanics approach. *Can. J. Civil Eng.*, **9**, 502–514

Keith, C.T. (1971) The anatomy of compression failure in relation to creep-inducing stress, *Wood Sci.*, **4** (2), 71–82.

Keith, C.T. (1972) The mechanical behaviour of wood in longitudinal compression, *Wood Sci.*, **4** (4), 234–244.

Koch, G., Schwab, E. and Kruse, K (1996) Investigation on shock resistance of wood from secondary damaged spruce (*Picea abies*) from wind-exposed high altitude stands in the Osterz mountains. *Holz als Roh. und Werkstoff*, **54** (5), 313–319.

Lavers, G.M. (1969/1983) *The strength properties of timber*, Bulletin 50, Forest Products Research Laboratory (2nd edn) HMSO. (3rd edn revised by G. Moore, 1983.)

Leicester, R.H. (1974) *Fracture strength of wood*, presented at First Australian Conference on Engineering Materials, University of New South Wales, Sydney.

Liu, J.Y., Zahn, J.J. and Schaffer, E.L. (1994) Reaction rate model for the fatigue strength of wood. *Wood and Fiber Sci.*, **26** (1), 3–10.

McDowell, B.J. (1984) Duration of load – a fracture mechanics approach tested experimentally, in *Proc. Timber Eng. Conf., Auckland, N. Z., vol III*: *Wood Science*, Ed. J. D. Hutchinson, Wellington, NZ.

McNatt, J.D. (1978) Linear regression of fatigue data, *Wood Sci.*, **11** (1), 39–41.

Madsen, B. (1973) *Duration of load tests for wet lumber in bending.* Structural Research Series Report 4, Dept. Civil Eng., University of British Columbia, Vancouver.

Madsen, B. (1979) Timestrength relationship for lumber, in: *Proc. First International Conference on Wood Fracture,* Banff, Alberta 1978, Ed. J.D. Barrett and R.O. Foschi, Forintek Canada Corp, pp. 111–128.

Madsen, B. (1982) Recommended moisture adjustment factors for lumber stresses. *Can. J. Civil Eng.*, **9**, 602–610.

Madsen, B. and Barrett, J.D. (1976) *Time-strength relationships for lumber*, Structural Research Series, Report No. 13, Dept. Civil Eng., University of British Columbia, Vancouver.

Mark, R.E. (1967) *Cell Wall Mechanics of Tracheids,* Yale University Press, New Haven.

Mindness, S., Nadeau, J.S. and Barrett, J.D. (1975) Slow crack growth in Douglas fir. *Wood Sci.*, **8** (1), 389–396.

Mindess, S., Barrett, J.D. and Spencer, R.A. (1980) Time-dependent failure of wood, in *Proc. Third International Conference on Mechanical Behaviour of Materials*, Cambridge, 1979, Ed. K J. Miller and R.F. Smith, Vol. **3**, pp. 319–328.

Mohr, B. (1996) Fatigue of structural timber, in *Proc. of International Conference on Wood Mechanics*, Stuttgart, Germany, Ed. S. Aicher, and published by the management committee of EC COST 508 Action, 217–225.

Moore, G.L. (1984) The effect of long-term temperature cycling on the strength of wood. *J. Inst. Wood Sci.*, **9** (6), 264–267.

Morlier, P., Valentin G and Toratti, T. (1994) Review of the theories on long term strength and time to failure, in *Proc. Workshop on Service Life Assessment of Wooden Structures*, Espoo, Finland, Ed. S.S. Gowda, and published by the management committee of EC COST 508 Action, 3–27.

Nadeau, J.S., Fuller, E.R. and Bennett, R. (1982) An explanation for the rate-of-loading and duration-of-load effects in wood in terms of fracture mechanics. *J. Mat. Sci.*, **17** (10), 2831–2840.

Nielsen, L. (1986) Wood as a cracked viscoelastic material, Part 1: Theory and application, pp 67–78. Part II: Sensitivity and Justification of a theory, pp 79–89, in *Proc. International Workshop* on *Duration of Load in Lumber and Wood Products* (1985) Vancouver, Canada, Spec. Pub. No. SP-27.

Nielsen, L.F. (1997) Lifetime and residual strength of wood subjected to static and variable load, in *Proc-Int Conf. IUFRO, S5.02 Timber Engineering*, Copenhagen, 1997.

Pearson, R.G. (1972) The effect of duration of load on the bending strength of wood. *Holzforschung*, **26** (4), 153–158.

Pearson, R.G. (1974) Application of fracture mechanics to the study of the tensile strength of structural lumber, *Holzforschung,* **28** (1), 11–19.

Pierce, C.B. (1976) The Weibull distribution and the determination of its parameters for application to timber strength data, *Building Research Establishment, Current Paper* 26/76.

van der Put, T.A.C.M. (1983) *Deformation and damage processes in wood.* Delft University Press, Delft, The Netherlands.

van der Put, T.A.C.M. (1993) Explanation of the failure criterion for wood, in *Proc. Workshop on Wood Plasticity and Damage*, Limerick, Ireland, Ed. C. Birkinshaw, P. Morlier and I. Seoane, and published by the management committee of EC COST 508 Action, 65–73.

Schniewind, A.P. (1967) Creep-rupture of Douglas-fir under cyclic environmental conditions. *Wood Sci. Technol.*, **1**, 278–288.

Schniewind, A.P. (1977) Fracture toughness and duration of load factor. II: Duration factor for cracks propagating perpendicular to the grain. *Wood & Fiber*, **9** (3), 216–276.

Schniewind, A.P. and Centeno, J.C. (1973) Fracture toughness and duration of load factor. 1. Six principal systems of crack propagation and the duration factor for cracks propagating parallel to grain, *Wood and Fiber*, **5** (2), 152–159.

Schniewind, A.P. and Lyon, D.E. (1973) A fracture mechanics approach to the tensile strength perpendicular to grain of dimension lumber. *Wood Sci. Technol.*, **7**, 45–59.

Smith, F.W. and Penney, D.T. (1980) Fracture mechanics of butt joints in laminated wood beams. *Wood Sci.*, **12** (4), 227–235.

Suzuki, S., Tamai, A., and Hirai, N. (1982) Effect of temperature on orthotropic properties of wood. Part III: Anisotropy in the LR-plane. *Mokuzai Gakkaishi*, **28** (7), 401–406.

Tang, R.C. (1994) Overview of duration-of-load research on lumber and wood composite panels in North America, in *Proc. Workshop on Service Life Assessment of Wooden Structures*, Espoo, Finland, Ed. S.S. Gowda, and published by the management committee of EC COST 508 Action, 171–205.

Thompson, R.J.H., Bonfield, P.W., Dinwoodie, J.M. and Ansell, M.P. (1996) Fatigue and creep in chipboard. Part 3: The effect of frequency, *Wood Sci. Technol.*, **30**, 293–305.

Toratti, T. (1992) *Creep of timber beams in a variable environment*. PhD thesis, Helsinki University of Technology, Finland.

Tsai, K.T. and Ansell, M.P. (1990) The fatigue properties of wood in flexure. *J. Mat. Sci.*, **25**, 865–878.

Turkulin, J. and Sell, J. (1997) *Structural and fractographic study on weathered wood.* Fotshungs-und Arbeitsbericht 115/36 Abteilung Holz, EMPA, Switzerland.

Walsh, P.F. (1972) Linear fracture mechanics in orthotropic materials, *Eng. Fract. Mech.*, **4**, 533–541.

Wilkins, A.P. and Ghali, M. (1987) Relationship between toughness, cell wall deformations and density in *Eucalyptus pilularis. Wood Sci. Technol.*, **21**, 219–226.

Wood, L.W. (1951) *Relation of strength of wood to duration of load*, Report No 1916, Forest Products Laboratory, Madison.

Ylinen, A. (1957) Zur Theorie der Danerstandfestigkeit des Holzes, *Holz als Roh und Werkstoff*, **15** (5), 213–215.

Chapter 8

Durability of timber

8.1 Introduction

Durability is a term that has different concepts for many people. It is defined here in the broadest possible sense to embrace the resistance of timber to attack from a whole series of agencies whether physical, chemical or biological in origin.

By far the most important are the biological agencies, the fungi and the insects, both of which can cause tremendous havoc given the right conditions. In the absence of fire, fungal or insect attack, timber is really remarkably resistant and timber structures will survive, indeed have survived, incredibly long periods of time, especially when it is appreciated that it is a natural organic material with which we are dealing. Examples of well-preserved timber items now over 2000 years old are to be seen in the Egyptian tombs. Both fungal and insect attack are described in Section 8.3 together with the important aspect of the natural durability of the timber.

Another important aspect of durability of timber is its reaction to fire and this is discussed in some detail in Section 8.4.

The effect of photochemical, chemical, thermal and mechanical actions are usually of secondary importance in determining durability; these are briefly considered in Section 8.2.

8.2 Chemical, physical, and mechanical agencies affecting durability and causing degradation

8.2.1 *Photochemical degradation*

On exposure to sunlight the colouration of the heartwood of most timbers will lighten, as is the case for mahogany, afrormosia and oak, although a few timbers will actually darken, such as Rhodesian teak. Indoors, the action of sunlight will be slow and the process will take several years; however, outdoors the change in colour is very rapid taking place in a matter of months and is generally regarded as an initial and very transient stage in the whole process of *weathering*.

In weathering the action not only of light energy (photochemical degradation), but also of rain and wind results in a complex degrading mechanism which renders the timber silvery-grey in appearance. More important is the loss of surface integrity, a process which has been quantified in terms of the residual tensile strength of thin strips of wood (Derbyshire and Miller, 1981; Derbyshire *et al.*, 1995). The loss in integrity embraces the degradation of both the lignin, primarily by the action of ultraviolet (UV) light, and the cellulose by shortening of the chain length, mainly by the action of energy from the visible part of the spectrum. Degradation results in erosion of the cell wall and in particular the pit aperture and torus. Fractography using scanning electron microscopy has revealed that the progression of degradation involves initially the development of brittleness and the reduction in stress transfer capabilities through lignin degradation, followed by reductions in microfibril strength resulting from cellulose degradation (Derbyshire *et al.*, 1995).

However, the same cell walls that are attacked act as an efficient filter for those cells below and the rate of erosion from the combined effects of UV, light and rain is very slow indeed. In the absence of fungi and insects the rate of removal of the surface by weathering is of the order of only 1 mm in every 20 years. Nevertheless, because of the continual threat of biological attack, it is unwise to leave most timbers completely unprotected from the weather. It should be appreciated that during weathering, the integrity of the surface layers is markedly reduced, thereby adversely affecting the performance of an applied surface coating. In order to effect good adhesion the weathered layers must first be removed (see Section 9.4).

8.2.2 Chemical degradation

As a general rule, timber is highly resistant to a large number of chemicals and its continued use for various types of tanks and containers, even in the face of competition from stainless steel, indicates that its resistance, certainly in terms of cost, is most attractive. Timber is far superior to cast iron and ordinary steel in its resistance to mild acids and for very many years timber was used as separators in lead–acid batteries. However, in its resistance to alkalis timber is inferior to iron and steel: dissolution of both the lignin and the hemicelluloses occurs under the action of even mild alkalis.

Iron salts are frequently very acidic and in the presence of moisture result in hydrolytic degradation of the timber; the softening and darkish-blue discolouration of timber in the vicinity of iron nails and bolts is due to this effect.

Timber used in boats is often subjected to the effects of chemical decay associated with the corrosion of metallic fastenings, a condition frequently referred to as *nail sickness*. This is basically an electrochemical effect, the rate of activity being controlled by oxygen availability. Areas of different polarity are set up, the salt water which has permeated the timber acts as the electrolytic bridge, while the wet timber assumes the role of the conductor. Permeability of the timber is therefore a significant factor and care must be exercised in the selection of timbers to use only impermeable species. Alkali will be produced at the cathodic surfaces which, as noted above, will cause the timber to become soft and spongy, impairing its ability to hold the fastenings. In the anodic areas ions pass into solution and form a soluble metallic salt with the negative ions of the electrolyte. As noted above in the case of iron fastenings, the iron salts so formed cause degrade of the timber resulting in considerable loss in its strength and marked discolouration.

8.2.3 *Thermal degradation*

Prolonged exposure of timber to elevated temperatures results in a reduction in strength and a very marked loss in toughness (impact resistance). Thus, timber heated at 120 °C for one month loses 10% of its strength; at 140 °C the same loss in strength occurs after only one week (Shafizadeh and Chin, 1977). Tests on three softwood timbers subjected to daily cycles of 20 °C to 90 °C for a period of three years resulted in a reduction in toughness to only 44% of its value of samples exposed for only one day (Moore, 1984). It has been suggested that degrade can occur at temperatures even as low as 65 °C when exposed for many years.

Thermal degradation results in a characteristic browning of the timber with associated caramel-like odour, indicative of burnt sugar. Initially this is the result of degrade of the hemicelluloses, but with time the cellulose is also affected with a reduction in chain length through scission of the β-1, 4 linkage; commensurate degrade occurs in the lignin, but usually at a slower rate (see Section 8.4).

8.2.4 *Mechanical degradation*

The most common type of mechanical degradation is that which occurs in timber when stressed under load for long periods of time. The concepts of duration of load (DOL) and creep are explained fully in Sections 7.6.12.2 and 6.3.1, respectively. Thus, Figures 7.12 and 7.13 illustrate how there is a loss in strength with time under load, such that after being loaded for 50 years the strength of timber is approximately only 50%. Similarly, there is a marked reduction in stiffness that manifests itself as an increase in extension or deformation with time under load, as illustrated in Figures 6.15 and 6.19. The structural engineer, in designing a timber structure, has to take into account the loss with time of both strength and stiffness by applying two time-modification factors.

A second and less common form of mechanical degradation is the induction of compression failure within the cell walls of timber, which can arise in the standing tree in the form of a *natural compression failure* due to high localised compressive stress, or as *brittleheart* due to the occurrence of high growth stresses (as described in Section 1.2.4), or under service conditions where the timber is overstressed in longitudinal compression with the production of kinks in the cell wall as described in Section 7.7.1.2. Loss in tensile strength due to the induction of compression damage is about 10–15%, but the loss in toughness can be as high as 50%. An example of failure in a scaffold board in bending resulting from the prior induction of compression damage due to malpractice on site is illustrated in Figure 8.1.

8.3 Natural durability and attack by fungi and insects

8.3.1 *Natural durability*

Generally when durability of timber is discussed reference is being made explicitly to the resistance of the timber to both fungal and insect attack; this resistance is termed *natural durability*.

In the UK, timbers have been classified into five durability groups which are defined in terms of the performance of the heartwood when buried in the ground. Examples of the more common timbers are presented in Table 8.1. Such an arbitrary classification is informative only in relative terms though these results

Figure 8.1 Kink bands and compression creases in the tension face of a scaffold board which failed in bending on site. Note how the crack pathway has followed the line of the top compression crease which had been induced some time previously; magnification × 150. (© BRE.)

on 50 × 50 mm ground stakes can be projected linearly for increased thicknesses. Timber used externally, but not in contact with the ground, will generally have a much longer life, though quantification of this is impossible.

Recalling that timber is an organic product it is surprising at first to find that it can withstand attack from fungi and insects for very long periods of time, certainly much greater than its herbaceous counterparts. This resistance can be explained in part by the basic constituents of the cell wall, and in part by the deposition of extractives (Sections 1.2.1 and 1.2.3.1; Table 1.2).

The presence of lignin that surrounds and protects the crystalline cellulose appears to offer a slight degree of resistance to fungal attack; certainly, the resistance of sapwood is higher than that of herbaceous plants. Fungal attack can commence only in the presence of moisture and the threshold value of 20% for timber is about twice as high as the corresponding value for non-lignified plants.

Timber has a low nitrogen content, of the order of 0.03–0.1% by mass and, as this element is a prerequisite for fungal growth, its presence in only such a small quantity contributes to the natural resistance of timber.

The principal factor conferring resistance to biological attack is undoubtedly the presence of extractives in the heartwood. The far higher durability of the heartwood of certain species compared with the sapwood is attributable primarily to the presence in the former of toxic substances, many of which are phenolic in origin. Other factors such as a decreased moisture content, reduced rate of diffusion, density and deposition of gums and resins also play a role in determining the higher durability of the heartwood.

Considerable variation in durability can occur within the heartwood zone: in a number of timbers the outer band of the heartwood has a higher resistance than the inner region owing, it is thought, to the progressive degradation of toxic substances by enzymatic or microbial action.

Durability of the heartwood varies considerably among the different species, being related to the type and quantity of the extractives present; the heartwood of timbers devoid of extractives has a very low durability. Sapwood of all timbers is susceptible to attack owing not only to the absence of extractives, but also the presence in the ray cells of stored starch which constitutes a ready source of food for the invading fungus.

The sapwood and heartwood of many species of timber can have its natural durability increased by impregnation with chemicals; the preservative treatment of timber is considered in Chapter 9.

8.3.2 Nature of fungal decay

In timber some fungi, such as the moulds, are present only on the surface and although they may cause staining they have no effect on the strength properties. A second group of fungi, the sapstain fungi, live on the sugars present in the ray cells and the presence of their hyphae in the sapwood imparts a

Table 8.1 Durability classification (resistance of heartwood to fungi in ground contact) for the more common timbers. (Note that the sapwood of all timber is perishable.)

	Classification (approximate life in ground contact)				
Timber type	*Perishable (< 5 years)*	*Non-durable (5–10 years)*	*Moderately durable (10–15 years)*	*Durable (15–25 years)*	*Very durable (> 25 years)*
Hardwoods	Alder Ash, European Balsa Beech, European Birch, European Horse chestnut Poplar, black Sycamore Willow	Afara Elm, English Oak, American red Obeche Poplar, grey Seraya, white	Avodire Keruing Mahogany, African Oak, Turkey Sapele Seraya, dark red Walnut, European Walnut, African	Agba Chestnut, sweet Idigbo Mahogany, American Oak, European Utile	Afrormosia Afzelia Ekki Greenheart Iroko Jarrah Kapur Makore Opepe Purpleheart Teak
Softwoods		Hemlock, western Parana pine Pine, Scots (redwood) Pine, yellow Podo Spruce, European (whitewood) Spruce, Sitka	Douglas fir Larch Pine, maritime	Pine, pitch Western red cedar Yew	

distinctive colouration to that region of the timber, when it is often referred to as 'blue-stain'. One of the best examples of sapstain is that found in recently felled Scots pine logs. In temperate countries the presence of this type of fungus results in only inappreciable losses in bending strength, though several staining fungi in the tropical countries cause considerable reductions in strength.

By far the most important group of fungi are those that cause decay of the timber by chemical decomposition; this is achieved by the digesting action of enzymes secreted by the fungal hyphae. Two main groups of wood-rotting fungi can be distinguished:

- *Brown rots* These consume the cellulose and hemicelluloses, but attack the lignin only slightly. During attack the wood usually darkens and in an advanced stage of attack tends to break up into cubes and crumbles under pressure. One of the best known fungi of this group is *Serpula lacrymans* which causes *dry rot*. Contrary to what its name suggests, the fungus requires an adequate supply of moisture for development.
- *White rots* These attack all the chemical constituents of the cell wall. Although the timber may darken initially, it becomes very much lighter than normal at advanced stages of attack. Unlike attack from the brown rots, timber with white rot does not crumble under pressure, but separates into a fibrous mass.

In very general terms, the brown rots are usually to be found in constructional timbers, whereas the white rots are frequently responsible for the decay of exterior joinery.

Decay, of course, results in a loss of strength, but it is important to note that considerable strength reductions may arise in the very early stages of attack; toughness is particularly sensitive to the presence of fungal attack. Loss in mass of the timber is also characteristic of attack and decayed timber can lose up to 80% of its air-dry mass.

The principal types of fungal attack of wood in the standing tree, of timber in felled logs, or of timber in service are set out in Table 8.2.

8.3.3 Nature of insect attack

Although all timbers are susceptible to attack by at least one species of insect, in practice only a small proportion of the timber in service actually becomes infested. Some timbers are more susceptible to attack than others and generally the heartwood is much more resistant than the sapwood. Nevertheless, the heartwood can be attacked by certain species particularly when decay is also present.

In certain insects the timber is consumed by the adult form and the best known example of this mode of attack are the *termites* where the adult, but sexually immature workers, cause most damage. Few timbers are immune to

Table 8.2 The principal types of fungal attack

Fungus	*Location of attack*			*Effect on the timber*	
	Tree	*Logs*	*Timber*	*Gross*	*Micro*
Brown rot	✓[a]	✓	✓	darkening of timber with cuboidal cracking	attacks cellulose and hemicellulose
White rot	✓[a]	✓	✓	bleaching of timber which turns fibrous	attacks cellulose, hemicellulose and lignin
Soft rot	–	–	✓	superficial; small cross-cracking; mostly occurring in ground contact	attacks cellulose of S_2 layer
Sapstain (blue-stain)	–	✓	✓	stains the sapwood of timber in depth	stain due to colour of hyphae
Moulds	–	✓	✓	superficial staining due to spores	live on cell contents; may increase permeability of timber
Bacteria	✓	✓	✓	subtle changes in texture and colour	increases permeability; get significant decay in ground contact

[a] Commonly as pocket rots.

attack by these voracious eaters and it is indeed fortunate that these insects generally cannot survive the cooler weather of this country. They are to be found principally in the tropics, but certain species are present in the Mediterranean region including southern France.

In this country insect attack is mainly by the grub or larval stage of certain beetles. The adult beetle lays its eggs on the surface of the timber, frequently in surface cracks, or in the cut ends of cells; these eggs hatch to produce grubs which tunnel their way into the timber, remaining there for periods of up to 3 years or longer. The size and shape of the tunnel, the type of detritus left behind in the tunnel (frass) and the exit holes made by the emerging adults are all characteristic of the species of beetle.

The principal types of insect attack – of wood in the standing tree, of timber in the form of felled logs, or of timber in service which may be free from decay or partially decayed – are set out in Table 8.3.

8.3.4 *Marine borers*

Timber used in salt water is subjected to attack by marine-boring animals such as the shipworm (*Teredo* sp.) and the gribble (*Limnoria* sp.). Marine borers are

Table 8.3 The principal types of insect attack of timber

Type of insect	*Location of attack*				*Comments*
	Tree	*Logs*	*Timber in service*		
			Sound	*Decayed*	
Pin-hole borers (Ambrosia beetle)	✓	✓	–	–	Produce galleries 1–2 mm diameter which are devoid of bore dust and are usually darkly stained; attack is frequently present in tropical hardwoods.
Forest longhorn	✓	(✓)	–	–	Galleries oval in cross-section; no bore dust but galleries may be plugged with coarse fibres; exit holes oval, 6–10 mm diameter.
Wood wasp	(✓)	✓	–	–	Galleries circular in cross-section and packed with bore dust; attacks softwoods; exit holes circular, 4–7mm diameter.
Bark borer beetle (*Ernobius mollis*)	(✓)	✓	(✓)	–	Requires presence of bark on timber; galleries empty and mainly in bark, but will also penetrate sapwood; exit holes circular, 1–2 mm diameter.
Powder-post beetle (*Lyctus*)	–	✓	✓	–	Attack confined to sapwood of hardwoods having large-diameter vessels; bore dust fine, talc-like; exit holes circular 1–2 mm diameter.
Common furniture beetle (*Anobium*)	–	–	✓	–	Attacks mainly the sapwood of both softwoods and European hardwoods; bore dust like lemon-shaped pellets; exit holes circular, 1–2 mm diameter.
House longhorn beetle-(*Hylotropes*)	–	–	✓	–	Attacks sapwood of softwoods mainly in the roof space of houses in certain parts of Surrey; bore dust sausage-shaped; exit holes few, oval, often ragged, 6–10 mm diameter.
Death-watch beetle (*Xestobium*)	–	–	(✓)	✓	Attacks both sap and heartwood of partially-decayed hardwoods, principally oak; bore dust bun-shaped; exit holes circular, 3 mm diameter.
Weevils (e.g. *Euophryum*)	–	–	–	✓	Attacks decayed softwoods and hardwoods in damp conditions; exit holes small, ragged about 1 mm diameter.
Wharf borer (*Narcerdes*)	–	–	–	✓	Attacks partially decayed timber to produce large galleries with 6 mm diameter oval exit hole.

particularly active in tropical waters; nevertheless, around the coast of Great Britain *Limnoria is* fairly active and *Teredo*, though spasmodic, has still to be considered a potential hazard. The degree of hazard will vary considerably with local conditions and there are relatively few timbers that are recognised as having heartwood resistant under all conditions; the list of resistant timbers includes ekki, greenheart, okan, opepe and pyinkado.

8.4 Performance of timber in fire

The performance of materials in fire is an aspect of durability that has always attracted much attention, not so much from the research scientist, but rather from the material user who has to conform with the legislation on safety and who is influenced by the weight of public opinion on the use of only 'safe' materials. Although various tests have been devised to assess the performance of materials in fire there is a fair degree of agreement in the unsatisfactory nature of many of these tests, and an awareness that certain materials can perform better in practice than is indicated by these tests.

Thus, while no one would doubt that timber is a combustible material showing up rather poorly in both the 'spread of flame' and 'heat release' tests, nevertheless in at least one aspect of performance, namely the maintenance of strength with increasing temperature and time, wood performs better than steel.

There is a critical surface temperature below which timber will not ignite. As the surface temperature increases above 100 °C, volatile gases begin to be emitted as thermal degradation slowly commences; however, it is not until the temperature is in excess of 250 °C that there is a sufficient build-up in these gases to cause ignition of the timber in the presence of a pilot flame. Where this is absent, the surface temperature can rise to about 500 °C before the gases become self-igniting. Ignition, however, is related not only to the absolute temperature, but also to the time of exposure at that temperature, since ignition is primarily a function of heat flux.

Generally chemical bonds begin to break down at about 175 °C and it is recognised that the first constituent of the timber to degrade is the lignin, and this continues slowly up to 500 °C. The hemicelluloses degrade much more quickly between 200 °C and 260 °C as does the cellulose within the temperature range 260–350 °C. Degradation of the cellulose results in the production of the flammable volatile gases and its marked reduction in degree of polymerisation (chain length; Le Van and Winandy, 1990).

The performance of timber at temperatures above ignition level is very similar to that of certain reinforced thermosetting resins which have been used as sacrificial protective coatings on space-return capsules. Both timber and these ablative polymers undergo thermal decomposition with subsequent removal of mass, leaving behind enough material to preserve structural integrity.

The onset of pyrolysis in timber is marked by a darkening of the timber and the commencement of emission of volatile gases; the reaction becomes exothermic and the timber reverts to a carbonised char popularly known as charcoal (Figure 8.2). The volatiles, in moving to the surface, cool the char and are subsequently ejected into the boundary layer where they block the incoming convective heat. This most important phenomenon is known as *transpirational cooling*. High surface temperatures are reached and some heat is rejected by thermal radiation. The heat balance is indicated in Figure 8.2. The surface layers crack badly both along and across the grain and surface material is continually, but slowly, being lost.

A quasi-steady state is reached, therefore, with a balance between the rate of loss of surface and the rate of recession of the undamaged wood. For most softwoods and medium-density hardwoods the rate at which the front recedes is about 0.64 mm/min, whereas for high-density hardwoods the value is about 0.5 mm/min (Hall and Jackman, 1975).

The formation of the char, therefore protects the unburnt timber which may be only a few millimetres from the surface. Failure of the beam or strut will occur only when the cross-sectional area of the unburnt core becomes too small to support the load. By increasing the dimensions of the timber above those required for structural consideration, it is possible to guarantee structural integrity in a fire for a given period of time. This is a much more desirable situation than that presented by steel, where total collapse of the beam or strut occurs at some critical temperature.

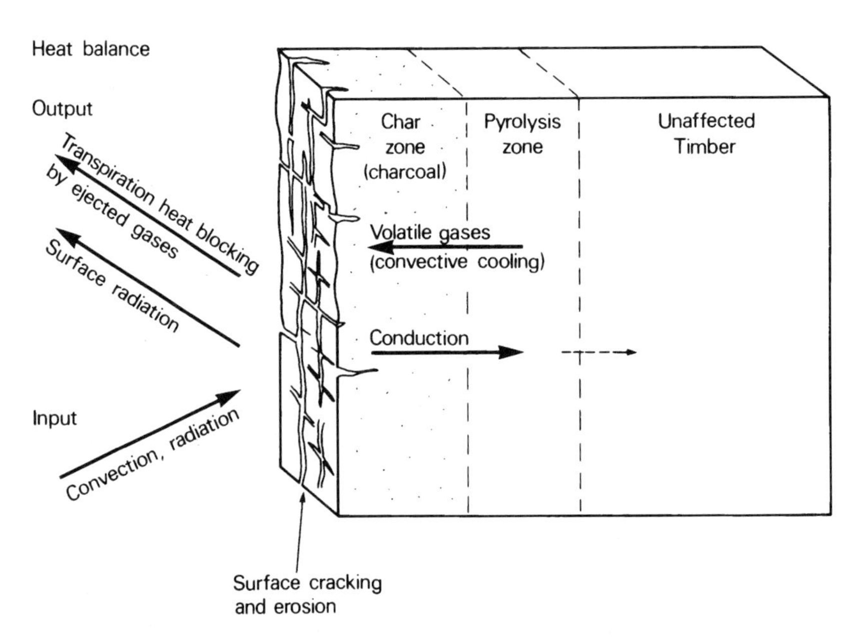

Figure 8.2 Diagrammatic representation of the thermal decomposition of timber. (© BRE.)

8.4.1 Methods of assessing performance

At the present time in the UK (1999), four standard tests are applicable to the evaluation of the performance of timber in fire, although this situation will change in the first few years of the new millennium with the introduction of the new European test methods to assess *reaction to fire*. The first of the existing standard tests is the *Non-combustibility Test for Materials* (BS 476 Part 4 (1970)) where a small sample of material is subjected to a temperature of 750°C; all timber and timber products, even when treated with fire retardants, are classified as *combustible*. The second test is a measure of ignitability where a small pilot flame is used to determine whether the sample will ignite easily (BS ISO 11925–2 (1997)); this standard recently replaced BS 476 Part 5 (1979), which rated timber and timber-based products as *not easily ignitable*.

Following ignition, the development of a fire is dependent on a number of factors, one of the more important being the rate of *spread of flame*. Using BS 476 Part 7 (1987) timber and wood-based panels (except cement-bonded particleboard (CBPB)) over 400 kg/m^3 are rated as Class 3, whereas timber products with a lower density are rated as Class 4. Cement-bonded particleboard contains a large mass of Portland cement and is rated Class 1.

For many applications, current regulations call for wall and ceiling linings to conform to Class 1; timber and timber products can be upgraded either by the application of intumescent paints to the surface, or by the incorporation of, or impregnation by, flame-retardant chemicals. These products influence the mechanism of decomposition, lower the temperature of onset of decomposition and increase the thickness of the char layer. The impregnation of timber by flame-retardant chemicals and coating it with intumescent paints are discussed in Sections 9.3.1.2 and 9.4, respectively.

The rate at which a combustible material contributes heat to a developing fire is a most important aspect and one in which timber and board materials do not show up very well, especially so when compared with alternative materials such as plasterboard. The *fire propagation test* (BS 476 Part 6 (1989)) provides some measure of the rate of heat release. Because of its large content of Portland cement, CBPB has a satisfactory rating of $I \leq 12$ and $i \leq 6$ and, as a result of it also having a Class 1 spread of flame (BS 476 Part 7 (1987)), this product is rated Class 0 in accordance with UK building regulations. (By way of explanation, I is the index of overall performance and characterises the heat release over the 20 min duration of the test, whereas i characterises the heat release over the first 3 min of the test. To be designated Class 0, a material must be (a) composed throughout of materials of limited combustibility or (b) be a Class 1 material that has a fire propagation index I of not more than 12 and a sub-index i of not more than 6.)

Thus, in three out of the four standard 'fire' tests currently carried out, timber and board products do not fair at all well. None of the four tests demonstrates the predictability of the performance of timber in fire, nor do they indicate the guaranteed structural integrity of the material for a calculable period of time.

The performance of timber in the widest sense is certainly superior to that indicated by the present set of standard tests.

Early in the first decade of the new millennium, National standards on reaction to fire will be replaced by new European standards. As far as the UK is concerned, the new European tests measure different aspects of the behaviour of building materials and products in fire than do BS 476 Parts 4–7. In the near future (all the new standards are currently present only in their final draft format), all construction products will be classified into one of six Euroclasses (A–F) according to their reaction-to-fire performance in fire tests. Two of these tests will be used to classify the least combustible materials (Euroclasses A_1 and A_2). These two new tests are a furnace test for *non-combustibility*, prEN ISO 1182 which is based on ISO 1182, but differing in small but significant detail, and an *oxygen bomb calorimeter test*, prEN ISO 1716 which is based on ISO 1716, but with modifications to improve consistency of operation.

At the lower end of the range of Euroclasses (classes E and F), construction products of appreciable combustibility will be assessed using a simple *ignitability* test, prEN ISO 11925–2. Products that fall into Classes A_2, B, C and D (and D will probably contain timber and wood-based panels) will be tested using the *single burning item test* (SBI) except where the products are used as floor coverings.

The SBI test requires the following determinations to be made:

- rate of heat release
- rate of smoke production
- rate of vertical and lateral flame spread
- time to ignition
- occurrence of flaming droplets or debris.

For floor coverings a *critical flux (radiant panel) test*, prEN ISO 9239–1, based on ISO 9239–1 will be used to determine performance in Euro classes B–E. For both floor and non-floor applications, generally only two of the above tests will be required to characterise performance of any one product.

The above reaction-to-fire tests relate to the product. When that product is incorporated into a building element, the *fire resistance* of that element will be determined by a whole series of other tests. Most of these are still at the drafting stage, but it is anticipated that there will be at least 60 European standards (or part-standards) relating to fire resistance of parts of a building.

References

Standards and specifications

BS 476 Part 4 (1970) and (1984) *Non-combustibility test for materials*, BSI, London.
BS 476 Part 5 (1979) *Method of test for ignitability*. BSI, London.
BS 476 Part 6 (1989) *Method of test for fire propagation for products*, BSI, London.

BS 476 Part 7 (1987) and (1993) *Method for classification of the surface spread of flame of products*, BSI, London.

BS ISO 11925–2 (1997) *Reaction to fire tests. Ignitability of building products subjected to direct impingement of flame. Part 2: Single flame source test*, BSI, London.

prEN ISO 1182 *Reaction to fire tests for building products – Non-combustibility test* (to be published, based on discussion document ISO/DIS 1182: 1998).

prEN ISO 1716 *Reaction to fire tests for building products – Determination of the gross caloric value* (to be published, based on discussion document ISO/DIS 1716: 1998).

prEN ISO 9239–1 *Reaction to fire tests for floor coverings – Part 1: determination of the burning behaviour using a radiant heat source* (to be published, based on discussion document ISO/DIS 9239–1: 1998).

prEN ISO 11925–2 *Reaction to fire tests for building products – Part 2: Ignitability when subjected to direct impingement of flame* (to be published, based on discussion document ISO/DIS 11925–2: 1998).

prEN ISO xxx *Reaction to fire tests for building products – The single burning item test*. (to be published)

Literature

Derbyshire, H. and Miller, E.R. (1981) The photodegradation of wood during solar radiation, *Holz als Roh-und Werkstoff*, **39**, 341–350.

Derbyshire, H., Miller, E.R., Sell, J. and Turkulin, H. (1995) Assessment of wood photodegradation by microtensile testing, *Drvna Ind.*, **46** (3), 123–132.

Hall, G.S. and Jackman, P.E. (1975). Performance of timber in fire, *Timber Trades J.*, 15 Nov, 38–40.

Le Van, S.L. and Winandy, J.E. (1990) Effects of fire retardant treatments on wood strength: a review. *Wood and Fiber Sci.*, **22** (1), 113–131.

Moore, G.L. (1984) The effect of long-term temperature cycling on the strength of wood, *J. Inst. Wood Sci.*, **9** (6), 264–267.

Shafizadeh, F. and Chin, P.P.S. (1977) Thermal degradation of wood, in, Goldstein, I. S. (Ed.) *Wood Technology: Chemical Aspects*, ACS Symposium Series 4, American Chemical Society, Washington, DC. pp. 57–81.

Chapter 9

Processing of timber

9.1 Introduction

After felling, the tree has to be processed in order to render the timber suitable for use. Such processing may be basically mechanical or chemical in nature or even a combination of both. On the one hand timber may be sawn or chipped, whereas on the other hand it can be treated with chemicals which markedly affect its structure and its properties. In some of these processing operations the timber has to be dried and this technique has already been discussed in Chapter 4 on water relationships and will not be referred to again in this chapter.

The many diverse mechanical and chemical processes for timber have been described in great detail in previous publications and readers desirous of such information are referred to the excellent and authoritative texts listed under Further Reading. In looking at processing in this chapter, the emphasis is placed on the properties of the timber as they influence or restrict the type of processing. For convenience, the processes are subdivided below into mechanical and chemical, but frequently their boundaries overlap.

9.2 Mechanical processing

9.2.1 Solid timber

9.2.1.1 Sawing and planing

The basic requirement of these processes is quite simply to produce as efficiently as possible timber of the required dimensions having a quality of surface commensurate with the intended use. Such a requirement depends not only on the basic properties of the timber, but also on the design and condition of the cutting tool; many of the variables are interrelated and it is frequently necessary to compromise in the selection of processing variables.

In Chapter 3 the density of timber was shown to vary by a factor of 10 from about 120 to 1200 kg/m^3. As density increases, so the time taken for the cutting edge to become blunt decreases. Whereas it is possible to cut over 10 000 ft of

Scots pine before it is necessary to resharpen, only 1000–2000 ft of a dense hardwood such as jarrah can be cut. Density will also have a marked effect on the amount of power consumed in processing. When all the other factors affecting power consumption are held constant, this variable is highly correlated with timber density, as illustrated in Figure 9.1.

Timber of high moisture content never machines as well as that at lower moisture levels. There is a tendency for the thin-walled cells to be deformed rather than cut because of their increased elasticity in the wet condition. After the cutters have passed over, these deformed areas slowly assume their previous shape resulting in an irregular appearance to the surface which is very noticeable when the timber is dried and painted; this induced defect is known as *raised grain*.

The cost of timber processing is determined primarily by the cost of tool maintenance, which in turn is related not only to properties of the timber, but also to the type and design of the saw or planer blade. In addition to the effect of timber density on tool life, the presence in certain timbers of gums and resins has an adverse effect because of the tendency for the gum to adhere to the tool thereby causing overheating. In saw blades this in turn leads to loss in tension resulting in saw instability and a reduction in sawing accuracy.

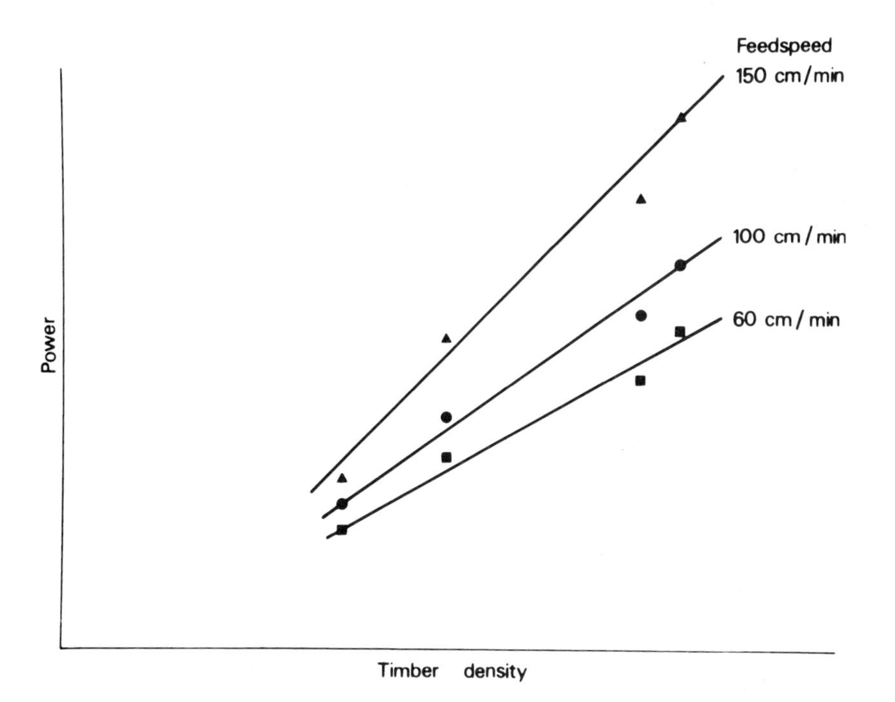

Figure 9.1 Effect of timber density and feedspeed on the consumption of power using a circular saw to cut along the grain (rip-sawing). (© BRE.)

A certain number of tropical hardwood timbers contain mineral inclusions which develop during the growth of the tree. The most common is silica which is present usually in the form of small grains within the ray cells; an example of a timber with a considerable amount of silica is given in Figure 9.2. The abrasive action of these inclusions is considerable and the life of the edge of the cutting tool is frequently reduced to almost one-hundredth of that obtained when cutting timber of the same density, but free of silica. Timbers containing silica are frequently avoided unless they possess special features which more than offset the difficulties which result from its presence.

One or two timbers are recognised as occasionally containing large deposits of calcium carbonate, a feature usually referred to as *stone*; the timber iroko is a well-known example. Not only stone, but also the presence in some timbers of nails, wire and even bullets does little to extend the life of expensive cutting tools.

The moisture content of the timber also plays a significant role in determining the life of cutting tools. As the moisture content decreases, so there is

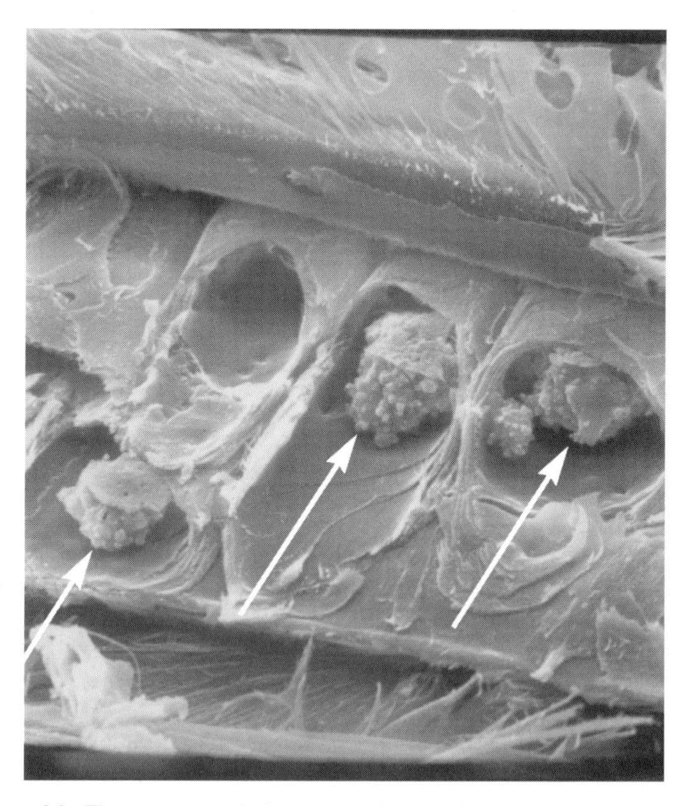

Figure 9.2 The presence of silica grains (arrowed) in the ray cells of *Parinari* species; scanning electron micrograph (× 15 000) (© BRE.)

a marked reduction in the time interval between resharpening both saw and planer blades. The fibrous nature of tension wood (Section 1.2.4) will also increase tool wear.

Service life will also depend on the type and design of the tool. Although considerably more expensive than steel, the use of tungsten-carbide-tipped saws and planer blades extends the life of the cutting edge, especially where timbers are either dense or abrasive. Increasing the number of teeth on the saw or the number of planer blades on the rotating stock will increase the quality of the surface, provided that the feedspeed is sufficient to provide a minimum bite per revolution; this ensures a cutting rather than a rubbing action which would accelerate blunting of the tool edge.

One of the most important tool design variables is the angle between the edge and the timber surface. As discussed in Section 1.2.2, timber is seldom straight-grained, tending in most cases to be in the form of a spiral of low pitch; occasionally, the grain is interlocked or wavy as discussed in Chapter 2. Under these circumstances, there is a strong tendency for those cells which are inclined towards the direction of the rotating cutter to be pulled out rather than cut cleanly, a phenomenon known as *pickup* or *tearing*. The occurrence of this defect can be removed almost completely by reducing the cutting angle (*rake angle*) of the rotating blades, though this will result in increased power consumption.

The cost of processing, though determined primarily by tool life, will be influenced also by the amount of power consumed. In addition to the effect of density of the timber previously discussed, the amount of energy required will depend on the feedspeed (Figure 9.1), tool design and, above all, on tool sharpness.

9.2.1.2 Steam bending

Steam bending of certain timbers is a long-established process which was used extensively when it was fashionable to have furniture with rounded lines. The backs of chairs and wooden hat stands are two common examples from the past, but the process is still employed at the present time, albeit on a much reduced volume. The handles for certain garden implements, walking sticks and a few sports goods are all produced by steam bending.

The mechanics of bending involves a presteaming operation to soften the lignin, swell the timber, and render the timber less stiff. With the ends restrained, the timber is usually bent round a former, and after bending the timber must be held in the restrained mode until it dries out and the bend is *set*. In broad terms the deformation is irreversible, but over a long period of time, especially with marked alternations in humidity of the atmosphere, a certain degree of recovery will arise, especially where the curve is unrestrained by some fixing. Although most timbers can be bent slightly, only certain species, principally the hardwood timbers of the temperate region, can be bent to sharp radii without cracking. When the timber is bent over a supporting but removable strap, the limiting radius of curvature is reduced appreciably. Thus, it is possible to bend 25 mm thick ash to a radius of 64 mm and walnut to a radius of only 25 mm. The significance of the anatomy of the timber in determining the limiting radius of curvature is still poorly understood.

9.2.2 Board materials

The area of board materials is recognised as being the fastest growing area within the timber industry since the late 1970s. Not only does this represent a greater volume of construction (particularly in the domestic area) and of consumer goods (such as furniture), but it also reflects a large degree of substitution of board materials for solid timber.

Production of wood-based panels in Europe in 1997 (the last year for which complete data are available) was 40.1×10^6 m^3, of which 72.5% was particleboard, 14% MDF, 7.5% plywood, 4.5% fibreboard and 1.5% OSB. Consumption would be in excess of production by about 6×10^6 m^3 (estimate) most of which can be attributed to large imports of plywood.

In the UK, the total consumption of board material in 1993 (the last year for which complete data are available) was 4.5×10^6 m^3 with a value in excess of £900 million; volume breakdown by panel products is given in Table 9.1.

Provisional data for consumption and production in the UK in 1997 are given in Table 9.2 from which it will be noted that the overall volume increase in consumption during 1993–1997 is about 30% with the increase in MDF being about 35%.

As a material, timber has a number of deficiencies:

- it possesses a high degree of variability;
- it is strongly anisotropic in both strength and moisture movement;
- it is dimensionally unstable in the presence of changing humidity;
- it is available in only limited widths.

Table 9.1 UK consumption of wood-based panels 1993

	Consumption (1993)	
Material	*Volume m^3 × 10^3*	*Proportion of total (%)*
Chipboard (particleboard)	2590	57
Plywood	1210	27
MDF	470	10
Hardboard	150	3
OSB	77	2
CBPB	4	1
Total	4501	100

Table 9.2 UK consumption and production of wood-based panels in 1997 (provisional figures)

Material	*Production* $m^3 \times 10^3$	*Imports* $m^3 \times 10^3$	*Exports* $m^3 \times 10^3$	*UK consumption* $m^3 \times 10^3$
Chipboard (particleboard) and OSB	2175	1180	240	3127
CBPB	11	9[a]	–	20[a]
MDF	412	316	88	631
Hardboard	–	250[a]	–	250[a]
Plywood	–	1800	–	1800[a]
Total	2598	3555	328	5828[a]

[a] = estimated data

Such material deficiencies can be lowered appreciably by reducing the timber to small units and subsequently reconstituting it, usually in the form of large flat sheets; moulded items are also produced, such as trays, bowls, coffins and chair backs. The degree to which these boards assume a higher dimensional stability and a lower level of anisotropy than is the case with solid timber is dependent on the size and orientation of the component pieces of timber and the method by which they are bonded together. There are an infinite variety of board types though there are only four principal ones – plywood, chipboard (particleboard), OSB and fibreboard.

In comparison with timber, board materials possess a lower degree of variability, lower anisotropy, and higher dimensional stability; they are also available in very large sizes. The reduction in variability is due quite simply to the random repositioning of variable components, the degree of reduction increasing as the size of the components decrease. The individual board materials are discussed below in separate sections.

9.2.2.1 *Plywood*

Over 1×10^6 m^3 of plywood are consumed annually in the UK. Most of this is made from softwood and is imported from the USA, Canada and Finland. Less than 1% of plywood is home produced.

Plywoods made from temperate hardwoods are imported mainly from Germany (beech) and Finland (birch – or birch/spruce combination) whereas plywoods from tropical hardwoods come predominately from south-east Asia (mainly Indonesia and Malaysia) and to a lesser, but increasing extent from South America and Africa.

Logs, the denser of which are softened by boiling in water, may be sliced into thin veneer for surface decoration by repeated horizontal or vertical cuts or, for plywood, peeled by rotation against a slowly advancing knife to give a continuous strip. After drying, sheets of veneer for plywood manufacture are coated with adhesive and are laid up and then pressed with the grain direction at right angles in alternate layers. Plywood frequently contains an unequal number of plies so that the system is balanced around the central veneer; some plywoods, however, contain an even number of plies, but with the two central plies having the same orientation, thereby ensuring that the plywood is balanced on each side of the central glue line.

As the number of plies increase, so the degree of anisotropy in both strength and movement drops quickly from the value of 40:1 for timber in the solid state. With three-ply construction and using veneers of equal thickness, the degree of anisotropy is reduced to 5:1, whereas for nine-ply material this drops to 1.5:1. However, cost increases markedly with number of plies and for most applications a three-ply construction is regarded as a good compromise between isotropy and cost.

The common multilayered plywood is technically known as a *veneer plywood* in contrast to the range of *core plywoods* where the surface veneers overlay a core of blocks or strips of woods.

Plywood (veneer type) for use in construction in Europe must comply with the requirements of EN 636, the three parts of which cover the use of the material in dry, humid or exterior conditions. The primary document supporting this specification is EN 314 (1993) on bond quality; Part 1 contains the actual test methods, and Part 2 sets out the requirements for each of the end-use climatic conditions. Bond performance depends primarily on the type of adhesive used, but this performance is also influenced by the amount of extenders used in its formulation as well as the quality of the veneers used.

The mechanical and physical properties of the plywood, therefore, will depend not only on the type of adhesive used, but also on the species of timber selected. Both softwoods and hardwoods within a density range of 400–700 kg/m^3 are normally utilised. Plywood for internal use is produced from the non-durable species and urea–formaldehyde (UF) adhesive. Plywood for external use is generally manufactured using phenol–formaldehyde (PF) resins; however, with the exception of marine grade plywood, durable timbers, or permeable non-durable timbers that have been preservative treated, are seldom used.

It is not possible to talk about strength properties of plywood in general terms because, not only are there different strength properties in different grain directions, but these are also affected by configuration of the plywood in terms of number, thickness, orientation and quality of the veneers and by the type of adhesive used. The factors that affect the strength of plywood are the same as those set out in Chapter 5 for the strength of timber, though the effects are not necessarily the same. Thus the intrinsic factors, such as knots and density, play a less significant part than they do in the case of timber, but the effects of the extrinsic variables such as moisture content, temperature and time are very similar to that for timber.

Plywood is the oldest of the timber sheet materials and for very many years has enjoyed a high reputation as a structural sheet material. Its use in the Mosquito aircraft and gliders in the 1940s, and its subsequent performance for small boat construction, for sheathing in timber-frame housing, and in the construction of web and hollow-box beams all bear testament to its suitability as a structural material.

When materials are compared in terms of their specific stiffness (stiffness per unit mass), plywood is stiffer than many other materials, including mild steel sheet; generally, plywood also has high specific strength. Another important property of plywood is its resistance to splitting, which permits nailing and screwing relatively close to the edges of the boards; this is a reflection of the removal of a line of cleavage along the grain which is a drawback of solid timber. Impact resistance (toughness) of plywood is very high and tests have shown that to initiate failure a force greater than the tensile strength of the timber species is required.

Plywoods tend to fall into three distinct groups. The first comprises those which are capable of being used structurally. Large quantities of softwood structural plywood are imported into the UK from North America, supplemented by smaller volumes from Sweden and Finland; the latter country also produces a birch–spruce structural plywood.

The use of this group of structural plywoods is controlled in that they must first comply with the European specification EN 636 (1997) and, second, the characteristic values for use in design must have been derived on semi-sized test pieces according to EN 789 (1996) and EN 1058 (1996). Because of the wide range of plywood types available, many of which are known by national trade names, each manufacturer of structural plywood must provide his own set of characteristic values. These must be accompanied by a certificate providing, in addition to other information, verification that these characteristic values have been obtained using EN 789 (1996) and EN 1058 (1996), that the plywood complies with EN 636 (1997), and that its production was under quality control and verified by a third party. Duration of load factors (K_{mod}) and creep factors (K_{def}) are included in Eurocode 5 (ENV 1995–1–1 (1994)).

The second group of plywoods comprises those which are used for decorative purposes, while the third group comprises those for general-purpose use. The latter are usually of very varied performance in terms of both bond quality and strength and are frequently used indoors for infill panels and certain types of furniture.

9.2.2.2 *Chipboard (particleboard)*

In the UK the boards made from wood chips and resin are known as chipboards; however, in most other countries the product is referred to as particleboard. The chipboard industry dates from the mid-1940s and originated with the purpose of utilising waste timber. After a long, slow start, when the quality of the board left much to be desired, the industry has grown tremendously since the late 1970s, far exceeding the supplies of waste timber available and now relying to a very large measure on the use of small trees for its raw material. Such a marked expansion is due in no small part to the much tighter control in processing and the ability to produce boards with a known and reproducible performance, frequently tailor-made for a specific end use. Over 60% of UK consumption is home-produced (see Table 9.2).

In the manufacture of chipboard the timber, which is principally softwood, is cut by a series of rotating knives to produce thin flakes which are dried and then sprayed with adhesive. Usually the chips are blown onto flat platens in such a way that the smaller chips end up on the surfaces of the board and the coarse chips in the centre. The mat is usually first cut to length before passing into the press where it is held for 0.10–0.20 min per millimetre of board thickness at temperatures up to 200 °C. The density of boards produced range from 450 to 750 kg/m^3, depending on end-use classification, whereas the resin content varies from about 9–11% on the outer layers to 5–7% in the centre layer, averaging out for the board at about 7–8% on a dry mass basis. Since the late 1980s many of the new chipboard plants have installed large continuous presses; as the name implies, the mat is fed in at one end to reappear at the other distant end as a fully cured board. This type of press has the advantage of being quick to respond to production changes in board thickness, adhesive type or board density.

Instead of using very long continuous press, chipboard can also be made continuously using either the Mendé or an extrusion process. The former is applicable only in the manufacture of thin chipboard, i.e. 6 mm or less, and the process is analogous to that of paper manufacture in that the board is sufficiently flexible to pass between and over large heated rollers. In the extrusion process, the chipboard mat is forced out through a heated die, but this results in the orientation of the chips at right angles to the plane of the board which reduces both the strength and stiffness of the material; it is used primarily as a core in the manufacture of doors and composite panels.

The performance of chipboard, like that of plywood, is very dependent on the type of adhesive used. Much of the chipboard (particleboard) produced in Europe is made using urea–formaldehyde (UF) which, because of its sensitivity to moisture, renders this type of chipboard unsuitable for use where there is a risk of the material becoming wet, or even being subjected to marked alternations in relative humidity over a long period of time. More expensive boards possessing some resistance to the presence of moisture are manufactured using either melamine-fortified urea formaldehyde (MUF) (see Figure 9.3), or phenol–formaldehyde (PF), or isocyanate (IS) adhesives; however, a true external-grade board has not yet been produced commercially.

Particleboard, like timber, is a viscoelastic material and an example of the deformation over an extended period of time has already been presented (Figure 6.15). However, the rate of creep in particleboard is considerably higher than

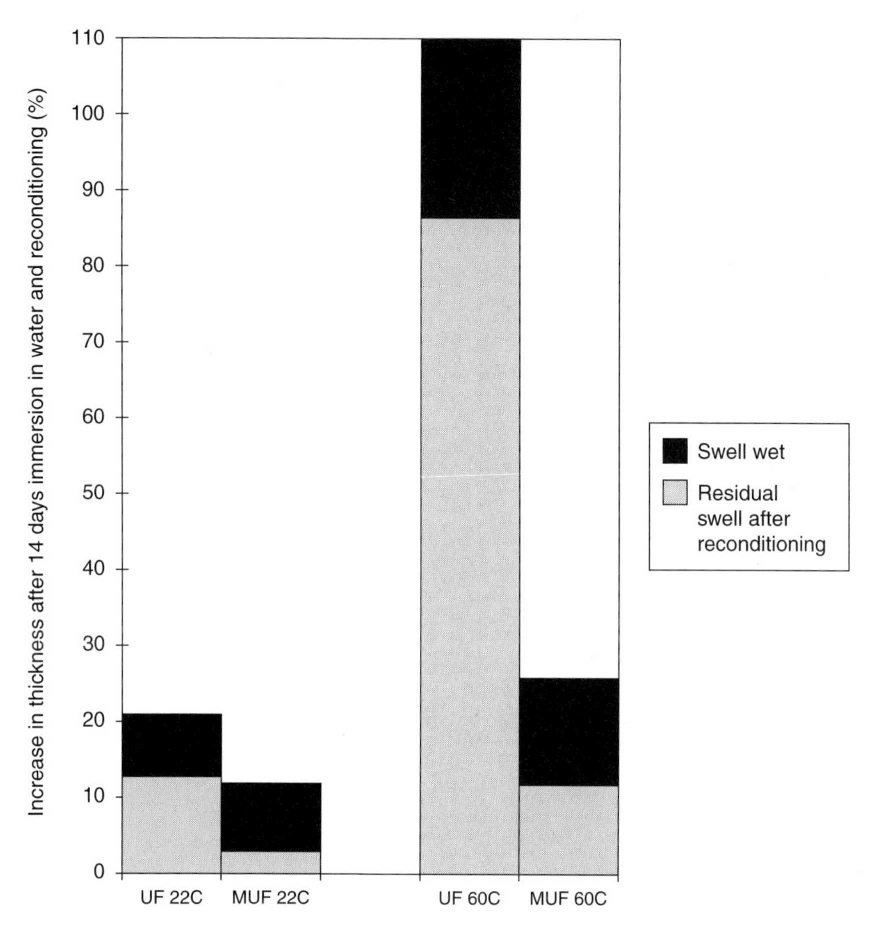

Figure 9.3 Comparison between UF- and MUF-bonded chipboard in the percentage increase in thickness following 14 days immersion in water at 22 °C and 60 °C, and in the residual increase in thickness following reconditioning at 20 °C and 65% relative humidity (© BRE.)

that in timber though it is possible to reduce it by increasing the amount of adhesive or by modifying the chemical composition of the adhesive.

Within the new framework of European specifications EN 312 sets out the requirements for six grades of particleboard of which four are rated as load bearing (i.e. they can be used in structural design). These are P4 and P6 for use in dry conditions, and P5 and P7 for use in humid conditions. The corresponding structural characteristic values for design use are presented in EN 12369 (to be published), whereas values of K_{mod} (the duration of load factor) and K_{def} (the creep factor) are to be found in Eurocode 5 (ENV 1995–1–1 (1994)).

Grades P2 and P3 are for non load-bearing use; the latter is a furniture/fittings board, whereas the former is a packaging and general purpose board. Particleboards are also produced from a wide variety of plant material and synthetic resin of which flaxboard and bagasse board are the best known examples.

9.2.2.3 *Wet-process fibreboard*

Although much smaller quantities of fibreboard are used in the UK than either chipboard or plywood (Table 9.2), it is nevertheless a most important panel product, used extensively in the UK for insulation and the linings of doors and backs of furniture, and in Scandinavia as a cladding and roofing material.

The process of manufacture is quite different from that of the other board materials in that the timber is first reduced to chips which are then steamed under very high pressure in order to soften the lignin which is thermoplastic in behaviour. The softened chips then pass to a defibrator which separates them into individual fibres, or fibre bundles without inducing too much damage.

The fibrous mass is usually mixed with hot water and formed into a mat on a wire mesh; the mat is then cut into lengths and, like particleboard, pressed in a multiplaten hot press at a temperature of from 180 °C to 210 °C; the board produced is smooth on only one side, the underside bearing the imprint of the wire mesh. By modifying the pressure applied in the final pressing, boards of a wide range of density are produced ranging from *softboard* with a density less than 400 kg/m^3, to *mediumboard* with a density range of 400–900 kg/m^3, to *hardboard* with a density exceeding 900 kg/m^3. Fibreboard, like the other board products, is moisture sensitive, but in the case of hardboard, a certain degree of resistance can be obtained by the passage of the material through a hot oil bath thereby imparting a high degree of water repellency; this material is referred to as *tempered* hardboard.

The European specification for these three groups of wet-process boards is EN 622; Parts 2, 3 and 4 cover, respectively, hardboard, mediumboard and softboard. Certain grades of hardboard (e.g. HB.HLA2 – a hardboard for heavy-duty load-bearing use under humid conditions) and mediumboard (MBH.LA2 – a high-density medium board for heavy-duty load-bearing use under dry conditions) are load-bearing boards for humid and dry conditions, respectively. The structural characteristic values to be used in design are set out in EN 12369; the duration of load factor (K_{mod}) and creep factor (K_{def}) are included in Eurocode 5 (ENV 1995–1–1 (1994)).

9.2.2.4 *MDF (Dry-process fibreboard)*

There has been a phenomenal increase in the production of dry-process fibreboard (MDF) world-wide over the last two decades. In the period from 1986 to 1997 European production rose from 0.58×10^6 m^3 to 5.5×10^6 m^3.

Consumption of MDF in the UK in 1997 was 0.6×10^6 m^3 of which 0.4×10^6 m^3 was home produced.

The fibre bundles are first dried to a low moisture content prior to being sprayed with an adhesive and formed into a mat which is hot-pressed to produce a board with two smooth faces similar to the production of particleboard; both multi-daylight and continuous presses are employed (see Section 9.2.2.2).

Various adhesive systems are employed. Where the board will be used in dry conditions a urea–formaldehyde (UF) resin is employed, whereas a board with improved resistance to moisture for use in humid conditions is usually manufactured using a melamine-fortified urea–formaldehyde resin (MUF), though phenol–formaldehyde (PF) or isocyanate (IS) resins are sometimes used.

The European specification for MDF is EN 622–5 (1997) which includes both load-bearing and non load-bearing grades for both dry and humid end uses; however, it should be appreciated that the load-bearing board for humid use (MDF.HLS – a medium-density fibreboard (i.e. MDF) for load-bearing use (short term only) under humid conditions) is restricted to only short-term periods of loading as defined in Eurocode 5. The structural characteristic values for design use for the load-bearing grades (MDF.LA – a medium-density fibreboard for load-bearing use under dry conditions – and MDF.HLS) are included in EN 12369, whereas the duration of load factor (K_{mod}) and creep factor (K_{def}) are set out in Eurocode 5 (ENV1995–1–1 (1994)).

A very large part of MDF production is taken up in the manufacture of furniture where non load-bearing grades for dry use are appropriate.

9.2.2.5 *OSB (Oriented strand board)*

Like MDF, OSB manufacture is fairly new and is certainly a growth area with over 50 mills world-wide with a capacity in 1997 in excess of 16×10^6 m^3; European capacity in 1997 was about 1×10^6 m^3; UK production from one mill in the same year was about 240 000 m^3.

Strands up to 75 mm in length with a maximum width of half its length are generally sprayed with an adhesive at a rate corresponding to about 2–3% of the dry mass of the strands. It is possible to work with much lower resin concentrations than with chipboard manufacture due to the removal of dust and 'fines' from the OSB line prior to resin application. In a few mills powdered resins are used, though most manufactures use a liquid resin. In the majority of mills a phenol–formaldehyde (PF) resin is used, but in one or two mills, a melamine-fortified urea formaldehyde (MUF) or isocyanate (IS) resin is employed.

In the formation of the mat the strands are aligned either in each of three layers, or only in the outer two layers of the board. The extent of orientation varies among manufacturers with property level ratios in the machine to cross-direction of 1.25:1 to 2.5:1, thereby emulating plywood. Indeed, the success of OSB has been as a cheaper replacement for plywood, but it must be appreciated that its strength and stiffness are considerably lower than those of high-quality structural grade plywood, though only marginally lower than those of many of the current structural softwood plywoods. It is widely used for suspended flooring, sheathing in timber-frame construction and flat roof decking.

OSB in Europe is manufactured to EN 300 (1997) which sets out the requirements for four grades, three of which are load-bearing, covering both dry and humid applications. Structural characteristic values have recently been determined and are included in EN 12369; the duration of load factors (K_{mod}) and creep factors (K_{def}) are present in Eurocode 5 (ENV 1995–1–1 (1994)).

9.2.2.6 Cement-bonded particleboard

Cement-bonded particleboard (CBPB) is very much a special end-use product manufactured in relatively small quantities. It comprises by mass 70–75% of Portland cement and 25–30% of wood chips similar to those used in particleboard manufacture. The board is heavy with a density of about 1200 kg/m^3, but it is very durable (due to its high pH of 11), is more dimensionally stable under changing relative humidity (due to the high cement content), has very good performance in reaction to fire tests (again because of the high cement content) and has poor sound transmission (due to the high density). The board is therefore used in high-hazard situations with respect to moisture, fire or sound.

One grade is specified in EN 634. The structural characteristic values are not available, but can be derived using EN 789 and 1058; duration of load and creep factors are also not available and must be determined according to ENV 1156.

9.2.2.7 Comparative performance of wood-based boards

With such a diverse range of board types, each manufactured in several grades, it is exceedingly difficult to select examples in order to make some form of comparative assessment.

In general terms, the strength properties of good quality structural softwood plywood are not only generally higher than all the other board materials, but they are usually similar to or slightly higher than that of softwood timber. In passing, it is interesting to note the reduction in anisotropy in bending strength from 4.5 for three-ply construction to 1.8 for seven-ply layup. Next to a good quality structural plywood in strength are the hardboards, followed by MDF and OSB. Chipboard (particleboard) is of lower strength, but still stronger than the mediumboards and CBPB. Table 9.3 provides the five percentile strength values included in the EN product specifications, with the exception of plywood where actual test data for Douglas fir plywood have been used. Other structural softwood plywoods can have strength values lower than those for Douglas fir, being similar to or only slightly above those of OSB of high quality. Actual strength values of individual manufacturer's products of non-plywood panels may be higher than these minimum specification values given in Table 9.3.

Table 9.3 Five percentile strength and stiffness and 95 percentile swell values for timber and board materials: quality control values in the EN product specifications

Material	Thickness (mm)	Density (kg/m^3)	EN 310 Bending strength (N/mm^2) Parallel	EN 310 Bending strength (N/mm^2) Perpendicular	EN 310 E (N/mm^2) Parallel	EN 310 E (N/mm^2) Perpendicular	EN 319 Internal bond (N/mm^2)	EN 317 24 h swelling thickness (%)	Moisture resistance Internal bond after Cyclic EN 321	Moisture resistance Internal bond after Boil EN 1087–1
Solid timber (Douglas fir)										
Small clear test pieces	20	590	80[b]	2.2[b]	16 400 [b]	1100[b]	–	–	–	–
Structural timber	100	580	22	–	8 110	–	–	–	–	–
Plywood (EN 636–3)										
Douglas fir (three ply)	4.8	520[a]	51[a]	11[a]	8 462[a]	624[a]	–	–	–	–
Douglas fir (seven ply)	19	600[a]	42[a]	23[a]	7 524[a]	2496[a]	–	–	–	–
OSB/3 (EN 300)	18	670[a]	20	10	3500	1400	0.32	15	0.15	0.13
OSB/4	18	670[a]	28	15	4800	1900	0.45	12	0.17	0.15
Particleboard (EN 312)										
Type P4 (Load-bearing dry)	15	720[a]	15		2150		0.35	15	–	–
Type P7 (Heavy load-bearing humid)	15	740[a]	20		3100		0.70	8	0.36	0.23
CBPB (EN 634)	18	1000	9		4000		0.50	1.5	0.3	–
Fibreboard (EN 622)										
Part 2 hardboard (load-bearing humid) HB.HLA1	3.2	900[a]	38		3800		0.80	15	–	0.5
Part 3 mediumboard (load-bearing dry) MBH.LA1	10	500[a]	18		1800		0.10	15	–	–
Part 5 MDF (load-bearing humid) MDF.HLS	12	790[a]	32		2800		0.80	10	0.25	0.15

[a] Not in the specifications.
[b] Mean values

It should be realised that these specification values are only for the purpose of quality control and must never be used in design calculations.

Comparison of the behaviour of these products to the effect of 24 h cold water soaking is also included in Table 9.3. CBPB is far superior to all other boards. Even higher swell values are to be found in 15 mm OSB/1 (general purpose board) of 25% and in 3.2 mm hardboard (HB.LA – load-bearing dry) of 35%. For those boards listed in the specifications for use under humid conditions that are included in Table 9.3, the table provides information on their moisture resistance in terms of their retention of internal bond strength following either the cyclic exposure test (EN 321), the boil test (EN1087–1) or both.

9.2.3 Laminated timber

The process of cutting timber up into strips and gluing them together again has three main attractions. Defects in the original piece of timber such as knots, splits, reaction wood, or sloping grain are redistributed randomly throughout the composite member making it more uniform in quality in comparison to the original piece of timber where the defects often result in stress raisers when load is applied. Consequently, the strength and stiffness of the laminated product will usually be higher than that of the timber from which it was made.

The second attraction is the ability to create curved beams or complex shapes, and the third is the ability to use shorter lengths of timber which can be end-jointed.

9.2.3.1 Glulam

This product has been around for many years and is to be found in the form of large curved beams in public buildings and sports halls. In manufacture, strips of timber about 20–30 mm in depth are coated with adhesive on their faces and laid up parallel to one another in a jig, the whole assembly being clamped until the adhesive has set. Generally, cold-setting adhesives are used because of the size of these beams; for dry end use a urea–formaldehyde (UF) resin is employed, whereas for humid conditions a resorcinol-formaldehyde (RF) resin is employed. The individual laminae are end jointed using either a scarf (sloping) or finger (interlocked) joint. Structural characteristic values for glulam are determined by the strength class of the timber(s) from which it is made, factored for the number and type(s) of laminates used.

9.2.3.2 Laminated veneer lumber

Laminated veneer lumber (LVL) is a product of the 1980s, one of three new products know as *engineered structural lumber* (see Sections 9.2.3.3 and

9.2.3.4). LVL is produced from softwood logs which are rotary peeled to produce veneers 3 mm in thickness. After kiln drying, these veneers are coated with a PF adhesive and bonded together under pressure to produce a board up to 24 m in length by 1.2 mm in width and up to 89 mm in thickness. This board is then cut up into structural timber battens. Characteristic values for design uses are available from the three manufacturers (two in Europe and one in Canada); these are from 50% to 100% higher than those of structural softwood timber and also possess much lower variability.

9.2.3.3 *Parallel strand lumber*

Parallel strand lumber (PSL) is a new North American product in which the 2.5 mm thick rotary peeled Douglas fir or Southern pine veneer is cut into strands some 2400 mm in length by 3 mm in width. These strands are coated with a resin, pressed together and microwave cured; battens up to 20 m in length can be produced.

9.2.3.4 *Laminated strand lumber*

Laminated strand lumber (LSL) is another new North American product in which Aspen veneer is cut into strands some 300 mm in length and 10 mm in width which are coated with an isocyanate resin before being aligned parallel to each other and pressed in a stream injection press to produce battens.

9.2.3.5 *Scrimber*

This is an Australian product in which green logs are passed through heavy steel rollers to crush the timber into a fibrous network. This is dried, coated with a PF resin and pressed into thick sheets which are cut up to produce battens.

9.2.4 Mechanical pulping

The pulping industry is the single largest consumer of wood. In the UK the consumption of paper and board in 1997 was 12.2×10^6 tonnes of which approximately half was home produced from home-grown softwood and hardwood, softwood residues, recycled fibre and imported woodpulp.

Pulp may be produced by either mechanical or chemical processes and it is the intention to postpone discussion on the latter until later in this chapter. In the original process for producing mechanical pulp, logs with a high moisture content are fed against a grinding wheel which is continuously sprayed with water in order to keep it cool and free it of the fibrous mass produced. The pulp so formed, known as stone groundwood, is coarse in texture, comprising bundles of cells rather than individual cells, and is mainly used as newsprint.

To avoid the necessity to adopt a costly bleaching process only light-coloured timbers are accepted. Furthermore, because the power consumed on grinding is a linear function of the timber density, only the low-density timbers with no or only small quantities of resin are used.

Much of the mechanical pulp now used is produced by disc-refining. Wood chips, softened in hot water, by steaming or by chemical pretreatment, are fed into the centre of two high-speed counter-rotating, ridged metal plates; on passing from the centre of the plates to the periphery the chips are reduced to fine bundles of cells or even individual cells. This process is capable of accepting a wider range of timbers than the traditional stone groundwood method.

9.3 Chemical processing

9.3.1 *Treatability*

The ease with which a timber can be impregnated with liquids, especially wood preservatives, is generally referred to as its treatability. Treatability is related directly to the permeability of timber which was discussed in some detail in Chapter 5 on flow. In that chapter the pathways of flow were described and it will be recalled that permeability was shown to be a function not only of moisture content and temperature, but also of grain direction, sapwood–heartwood, earlywood–latewood and species.

Longitudinal permeability is usually about 10 000 times greater than transverse permeability owing principally to the orientation of the cells in the longitudinal direction. Heartwood, owing to the deposition of both gums and encrusting materials, is generally much less permeable than the sapwood, whereas earlywood of the sapwood in the dry condition has a much lower permeability than the same tissue in the green state due to aspiration of the bordered pits in the dry state.

Perhaps the greatest variability in ease of impregnation occurs between species. Within the softwoods this can be related to the number and distribution of the bordered pits and to the efficiency of the *residual* flow paths which utilise both the latewood bordered pits and the semibordered ray pits. Within the hardwoods variability in impregnation is related to the size and distribution of the vessels and to the degree of dissolution of the end walls of the vessel members.

Four arbitrary classes of treatability are recognised, different timbers being assigned to these according to the depth and pattern of penetration when treated with wood preservatives (see EN 350–2). These classes are *permeable*, *moderately resistant*, *resistant* and *extremely resistant*. This classification is equally applicable to impregnation by flame retardants or dimensional stabilisers because, although differences in viscosity will influence degree of penetration, the treatability of the timber species will remain in the same relative order.

9.3.1.1 Artificial preservatives

Except where the heartwood of a naturally durable timber is being used, timber should always be treated with a wood preservative if there is any significant risk that its moisture content will rise above 20% during its service life. At and above this moisture content, wood-destroying fungi can attack. The relationship between service environment and risk of attack by wood-destroying organisms is defined in EN 335–1 using the hazard classification of biological attack. Clearly, those timbers of greater permeability will take up preservatives more easily and are to be preferred over those that are more resistant. It is normally not necessary to protect internal woodwork which should remain dry. However, where the risk of water spillage, leakage from pipes or from the roof is seen as likely or significant, application of wood preservatives may be deemed a sensible precaution.

A variety of methods for the application of wood preservatives are available. Short-term dipping and surface treatments by brush or spray are the least effective ways of applying a preservative because of the small loading and poor penetration achieved. In these treatments only the surface layers are penetrated and there is a risk of splits occurring during service which will expose untreated timber to the risk of attack by wood-destroying organisms. Such treatments are usually confined to do-it-yourself treatments, or treatments carried out during remediation or maintenance of existing woodwork.

The most effective methods of timber impregnation are industrial methods in which changes in applied pressure ensure controlled, more uniform penetration and retention of preservative. The magnitude of the pressure difference depends on the type of preservative being used. Essentially, the timber to be treated is sealed in a pressure vessel and a vacuum drawn. While under vacuum, the vessel is filled with the preservative and then returned to atmospheric pressure during which some preservative enters the wood. At this point, an over-pressure of between zero and 13 bar is applied, depending mainly on the preservative being used, but also on the treatability of the timber. This can be held for between several minutes and many hours after which the vessel is drained of preservative. A final vacuum is often applied to recover some of the preservative and to ensure the treated timber is free of excess fluid. Details of the choice, use and application of wood preservatives can be found in BS 1282 (1975).

There are three main types of preservative in general use. The first group are the tar oils of which coal tar creosote is the most important. Its efficacy as a preservative lies not only in its natural toxicity, but also in its water-repellency properties. It has a very distinctive and heavy odour and treated timber cannot be painted unless first coated with a metallic primer.

The second group are the waterborne preservatives which are suitable for both indoor and outdoor uses. The most common formulations are those containing copper, chromium and arsenic compounds, but combinations of copper–chromium and copper–chromium–boron are also used. All of these preservatives are usually applied by a vacuum high-pressure treatment; the chemicals react once in the wood and become *fixed*, i.e. they are not leached out in service.

Inorganic boron compounds are also used as water-borne preservatives, but they do not become fixed within the wood and therefore can be leached from the wood during service. Their use is therefore confined to environments where leaching cannot take place.

The third group are the solvent-type preservatives, which tend to be more expensive than those of the first two groups, but they have the advantage that machined timber can be treated without the grain being raised, as would be the case with aqueous solutions. The formulations of the solvent type are based on a variety of compounds including pentachlorophenol, tri-*n*-butyltin oxide, and copper and zinc naphthanates. More recently, copper and zinc versatate, zinc ocoate, and the preparations known as acypetacs zinc and acypetacs copper have been introduced. Some organic solvent preservatives include insecticides and water repellents. BS 5707 lists the standardised formulations. These preservatives find uses in the do-it-yourself and industrial sectors. Industrial treatment processes include double vacuum and immersion techniques, whereas do-it-yourself and on-site treatment includes dipping and brushing.

In looking at the application of these three different types of preservatives, creosote and CCA (copper-chromium-arsenic) are able to protect timber in high-hazard situations, namely ground contact, whereas organic solvent preservatives are used in timber out of ground contact and preferably protected with a paint film.

Although it is not an impregnation process as defined above, it is convenient to examine here the diffusion process of preservation. The timber must be in the green state and the preservative must be water soluble. Timber is immersed for a short period in a concentrated (sometimes hot) solution of a boron compound, usually disodium octoborate tetrahydrate, and then close-stacked undercover for several weeks to allow the preservative to diffuse into the timber.

Gaseous diffusion has been examined as a method of preservative treating wood-based panels as well as timber. Treatment by trimethyl borate vapour in a sealed, evacuated chamber of material at a low moisture content results in the reaction of the gaseous trimethyl borane with the water in the wood to produce the biocide boric acid which is then deposited where the reaction occurs (Turner and Murphy, 1998).

For many years there was some difference of opinion as to whether these preservatives merely lined the walls of the cell cavity or actually entered the cell wall. However, it has been well demonstrated by electron microanalysis that, whereas creosote only coats the cell walls the waterborne preservatives do impregnate the cell wall. However, because of the capillary nature of the cell wall, it was thought for some time that selective filtration might occur owing to the disparity in sizes of the constituent chemical groups present. Such doubts, however, were removed by the results of work using an electron microscope microanalyser, EMMA 4, fitted with a probe only 0.2 μm in diameter which allowed the distribution of copper, chromium and arsenic across the cell wall following impregnation to be obtained (Chou *et al.*, 1973). All three elements were present in all the regions examined and the distribution of this fine deposit is dictated by the distribution of the microfibrils; these appear to be coated individually with a layer some 1.5–2.0 nm in thickness. In some parts of the cell wall a coarser deposit is also present. The preservative was also found to be present on the surface of the cell cavity with concentrations of the elements some 2–5 times those within the cell wall.

Although there is some evidence to show that PCP (pentachlorophenol) can penetrate the cell wall, it is doubtful whether solvent-type preservatives *in general* penetrate the wall due to their large molecular size and by being carried in a non-polar solvent.

There is considerable variation in preservative distribution in treated dry timber; in the softwoods only the latewood tends to be treated due to aspiration of the earlywood bordered pits as described in Section 5.2.1.4. In the hardwoods, treatment is usually restricted to the vessels and tissue in close proximity to the vessels, again as described in Section 5.2.1.4.

In those timbers which can be impregnated, it is likely that the durability of the sapwood after pressure impregnation will be greater than the natural durability of the heartwood, and it is not unknown to find telegraph and transmission poles the heartwood of which is decayed while the treated sapwood is perfectly sound.

Mention has been made already of the difficulty of painting timber which has been treated with creosote. This disadvantage is not common to the other preservatives and not only is it possible to paint the treated timber, but it is also possible to glue together treated components.

9.3.1.2 Flame retardants

Flame-retardant chemicals may be applied as surface coatings or by vacuum–pressure impregnation, thereby rendering the timber less easily ignitable and reducing the rate of flame spread. Intumescent coatings will be discussed later and this section is devoted to the application of fire retardants by impregnation.

The salts most commonly employed in the UK for the vacuum pressure impregnation process are monoammonium phosphate, diammonium phosphate, ammonium sulphate, boric acid and borax. These chemicals vary considerably in solubility, hygroscopicity and effectiveness against fire. Most proprietary flame retardants are mixtures of such chemicals formulated to give the best performance at reasonable cost. As these chemicals are applied in an aqueous solution, it means that a combined waterborne preservative and fire retardant solution can be used which has distinct economic considerations. Quite frequently, corrosion inhibitors are incorporated where the timber is to be joined by metal connectors. Treatment of the timber involves high pressure–vacuum processes; as aqueous solutions are involved redrying of the timber is required.

Considerable caution has to be exercised in determining the level of heating to be used in drying the timber following impregnation. The ammonium phosphates and sulphate tend to break down on heating giving off ammonia and leaving an acidic residue which can result in degradation of the wood substance as described in Chapter 8. Thus, it has been found that drying at 65 °C following impregnation by solutions of these salts results in a loss of bending strength of from 10% to 30%. Drying at 90 °C, which is adopted in certain kiln schedules, results in a loss of 50% of the strength and even higher losses are recorded for the impact resistance or toughness of the timber. It is essential, therefore, to dry the timber at as low a temperature as possible and also to ensure that the timber in service is never subjected to elevated temperatures which would initiate or continue the process of acidic degradation. Most certainly, timber which has to withstand suddenly applied loads should not be treated with this type of fire retardants, and care must also be exercised in the selection of glues for construction. The best overall performance from timber treated with these flame retardants is obtained when the component is installed and maintained under cool and dry conditions.

Conscious of the limitations of flame retardants based on ammonium salts, a number of companies have developed effective retardants of very different chemical composition; these have much reduced degradation of the timber, or do not degrade the timber, and are usually non-corrosive to metal fixings. However, they are considerably more expensive than those based on ammonium salts.

9.3.1.3 Dimensional stabilisers

In Chapter 4 on movement, timber, because of its hygroscopic nature, was shown to change in dimensions as its moisture content varied in order to come into equilibrium with the vapour pressure of the atmosphere. Because of the composite nature of timber such movement will differ in extent in the three principal axes.

Movement is the result of water adsorption or desorption by the hydroxyl groups present in all the matrix constituents. Thus it should be possible to reduce movement (i.e. increase the dimensional stability) by eliminating or at least reducing the accessibility of these groups to water. This can be achieved by either chemical changes or by the introduction of physical bulking agents.

Dimensional stability is imparted to wood by swelling of the substrate due to chemical modification, as the bonded groups occupy volume within the cell wall. At high levels of modification, wood is swollen to near its green volume, and antishrink efficiencies close to 100% are achieved. After extended reaction, swelling in excess of the green volume can occur, which is accompanied by cell wall splitting.

Various attempts have been made to substitute the hydroxyl groups chemically by less polar groups and the most successful has been by acetylation (Rowell, 1984). In this process acetic anhydride is used as a source of acetyl groups. A very marked improvement in dimensional stability is achieved with only a marginal loss in strength. Using carboxylic acid anhydrides of varying chain length, Hill and Jones (1996) obtained good dimensional stabilisation which they attributed solely to the bulking effect. Other reagents with potential for chemical modification include isocyanates and epoxides. More information on chemical modification is available in the review by Rowell (1984) and the text edited by Hon (1996).

Good stabilisation can also be achieved by reacting the wood with formaldehyde which then forms methylene bridges between adjacent hydroxyl groups. However, the acid catalyst necessary for the process causes acidic degradation of the timber.

In contrast to the above means of chemical modification, a variety of chemicals have been used to physically stabilise the cell wall; these impregnants act as bulking agents and hold the timber in a swollen condition even after water is removed, thus minimising dimensional movement.

Starting in the mid-1940s and continuing on a modest scale to the present time, some solid timber, but more usually wood veneers, are impregnated with solutions of phenol–formaldehyde. The veneers are stacked, heated and compressed to form a high-density material with good dimensional stability which still finds wide usage as a heavy-duty insulant in the electrical distribution industry.

Considerable success has also been achieved using polyethylene glycol (PEG), a wax-like solid which is soluble in water. Under controlled conditions, it is possible to replace all the water in timber by PEG by a diffusion process, thereby maintaining it in a swollen condition. The technique has found application, among other things, in the preservation of waterlogged objects of archaeological interest, the best example of which is the Swedish wooden warship *Wasa*, which was raised from the depths of Stockholm harbour in 1961 having foundered in 1628. From 1961 the timber was sprayed continuously for over a decade with an aqueous solution of PEG which diffused into the wet timber, gradually replacing the bound water in the cell wall without causing any dimensional changes.

PEG may also be applied to dry timber by standard vacuum impregnation using solution strengths of from 5% to 30%. Frequently, preservative and/or fire-retardant chemicals are also incorporated in the impregnating solution. It will be noted from Figure 9.4 that, following impregnation, the amount of swelling has been reduced to one third that of the untreated timber.

It is also possible to improve the stability of timber by impregnating it with low-viscosity liquid polymers which can then be polymerised within the timber to convert them to solid polymers. Timber so treated is referred to as *polymer-impregnated wood* or *wood–plastic composite* (WPC). A whole range of

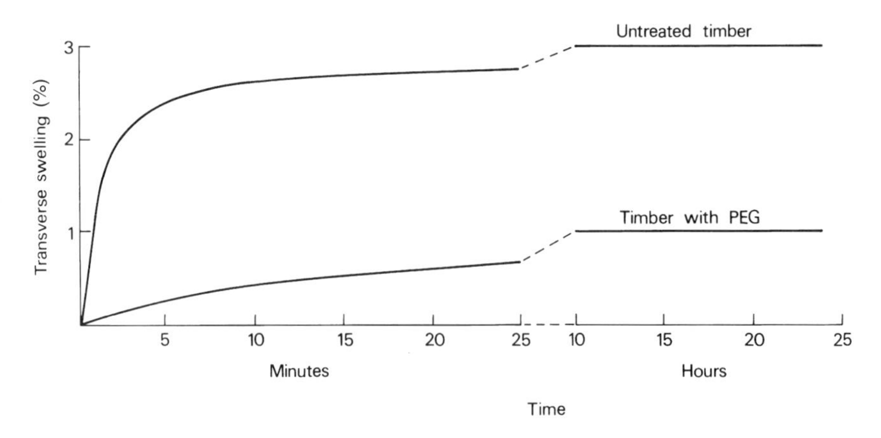

Figure 9.4 The comparative rates of swelling in water of untreated pine timber and timber impregnated with a 50% (by mass) solution of PEG (polyethylene glycol): this is equivalent to 22% loading on a dry wood basis. (Adapted from Morén (1964).)

monomers have been tried, but most success has been achieved with methyl methacrylate, or a (60:40) mixture of styrene and acrylonitrile.

After the timber is impregnated with the liquid monomer using standard vacuum–pressure impregnation techniques, the monomer is polymerised by either gamma irradiation or by the use of free-radical catalysts and heat. The former method is usually selected because a greater bulk of timber can be treated than is the case with heating; however, special equipment is required and stringent safety precautions must be enforced. The radiation dose must be controlled carefully to avoid degrade of the timber.

Swelling of the timber occurs during impregnation, which is indicative of the penetration of the cell wall by the monomer; after polymerisation of the monomer the timber possesses enhanced dimensional stability (Figure 9.5). The degree of penetration, and consequently the amount of dimensional stability, can be increased by adding swelling solvents, such as dioxan, to the impregnant.

It is possible that the polymer is acting in more than a bulking role in that it could form copolymers with the various chemical constituents of the cell wall. A limited amount of evidence is available to support this hypothesis of *grafting*, though some workers have ruled out such a possibility. Not only is the dimensional stability improved by this process, but a number of the strength properties are also increased. Modulus of rupture and compression strength parallel to the grain are usually increased slightly, while the hardness is raised almost threefold. However, the impact resistance and elasticity tend to fall slightly (Table 9.4).

WPC has a most attractive appearance, but unfortunately tends to cost some three to four times as much as untreated timber. Consequently it has been used

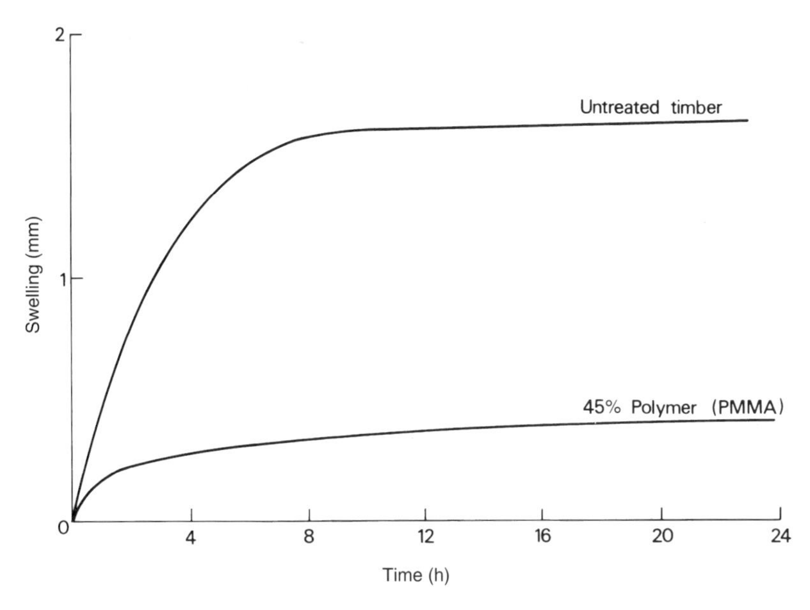

Figure 9.5 The comparative rates of swelling in water of untreated birch and a composite of the same wood with poly(methyl methacrylate) (PMMA). (© BRE.)

Table 9.4 Values of strength and stiffness of wood–plastic composites as a percentage of untreated timber

Composite[a]	*MOR (%)*	*Impact (%)*	*Compression along grain (%)*	*Radial hardness (%)*	*MOE (bending) (%)*
Birch + 45% MMA	116	68	119	228	94
Podo + 70% MMA	111	100	116	371	89

[a] MMA = methyl methacrylate on v/v basis; radiation dose = 4 Mrad.

for only small and specialised items such as cutlery handles and brush backs. The material has much potential as a flooring material for heavyduty areas such as dance halls, but so far, because of its cost, it has found little acceptance.

Developments in the production and use of water-repellent preservatives based on resins dissolved in low-viscosity organic solvents have resulted in the ability to confer on timber a low, but none the less important level of dimensional stability. Their application is of considerable proved practical significance in the protection of joinery out-of-doors and these are discussed further in Section 9.4.

9.3.2 Chemical pulping

The magnitude of the pulping industry has already been discussed as has the production of mechanical pulp. Where paper of a higher quality than newsprint or corrugated paper is required, a pulp must be produced consisting of individual cells rather than fibre bundles. To obtain this type of pulp the middle lamella has to be removed and this can be achieved only by chemical means.

There are a number of chemical processes which are described in detail in the literature. All are concerned with the removal of lignin, which is the principal constituent of the middle lamella. However, during the pulping process lignin will also be removed from within the cell wall as well as from between the cells. This is both acceptable and desirous as lignin imparts a greyish colouration to the pulp which is unacceptable for the production of white paper.

However, it is not possible to remove all the lignin without also dissolving most of the hemicelluloses which not only add to the mass of pulp produced, but also impart a measure of adhesion between the fibres. Thus, a compromise has to be reached in determining how far to progress with the chemical reaction and the decision is dependent on the requirements of the end product. Frequently, though not always, the initial pulping process is terminated when a quarter to a half of the lignin still remains and this is then removed in a subsequent chemical operation known as bleaching which, though expensive, has relatively little effect on the hemicelluloses. The yield of chemical pulp will vary considerably depending on the conditions employed, but it will usually be within the range of 40–50% of the dry mass of the original timber.

The yield of pulp can be increased to 55–80% by semichemical pulping. Only part of the lignin is removed in an initial chemical treatment which is designed to soften the wood chips; subsequent mechanical treatment separates the fibres without undue damage. These high-yield pulps usually find their way into card and board-liner which are extensively used for packaging where ultimate whiteness is not a prerequisite.

9.3.3 Other chemical processes

Brief mention must be made of the *destructive distillation* of timber, a process which is carried out either for the production of charcoal alone or for the additional recovery of the volatile by-products such as methanol, acetic acid, acetone and wood-tar. The timber is heated initially to 250 °C, after which the process is exothermic; distillation must be carried out either in the complete absence of air, or with controlled small amounts of air.

Timber can be softened in the presence of ammonia vapour as a result of plasticisation of the lignin. The timber can then be bent or moulded using this process but, because of the harmful effects of the vapour, the process has never been adopted commercially.

9.4 Finishes

Finishes have a combined decorative and protective function. Indoors they are employed primarily for aesthetic reasons, though their role in resisting soiling and abrasion is also important; outdoors, however, their protective function is vital. In Chapter 8, the natural weathering process of timber was described in terms of the attack of the cell wall constituents by ultraviolet light and the subsequent removal of breakdown products by rain; the application of finishes is to slow down this weathering process to an acceptable level, the degree of success varying considerably among the wide range of finishes commercially available.

In Chapter 1 the complex chemical and morphological structure of timber was described; in Chapter 4 the hygroscopic nature of this fibre composite and its significance in determining the movement of timber was discussed. The combined effects of structure and moisture movement have a most profound effect on the performance of coatings. For example in the softwoods the presence of distinct bands of early and latewood with their differential degree of permeability results not only in a difference in sheen or reflectance of the coating between these zones, but also in marked differences in adhesion; in Douglas fir, where the latewood is most conspicuous, flaking of paint from the latewood is a common occurrence. In addition, the radial movement of the latewood has been shown to be as high as six times that of the earlywood, and consequently the ingress of water to the surface layers results in differential movement and considerable stressing of the coatings. In those hardwoods characterised by the presence of large vessels, the coating tends to sag across the vessel and it is therefore essential to apply a paste filler to the surface prior to painting; even with this, the life of a paint film on a timber such as oak (see Figure 1.6) is very short. For this reason, the use of exterior wood stains (see later) is common, as this type of finish tends not to exhibit the same degree of flaking. The presence of extractives in certain timbers (see Section 1.2.3.1 and Table 1.2) results in the inhibition in drying of most finishes; with iroko and Rhodesian teak, many types of finish may never dry.

Contrary to general belief, deep penetration of the timber is not necessary for good adhesion, but it is absolutely essential that the weathered cells on the surface are removed prior to repainting. Good adhesion appears to be achieved by molecular attraction rather than by mechanical keying into the cell structure.

Although aesthetically most pleasing, fully exposed varnish, irrespective of chemical composition, has a life of only a very few years, principally because of the tendency of most types to become brittle on exposure, thereby cracking and disintegrating because of the stresses imposed by the movement of the timber under changes in moisture content. Ultraviolet light can readily pass through the majority of varnish films, degrading the timber at the interface and causing adhesion failure of the coating.

A second type of natural finish which overcomes some of the drawbacks of clear varnish is the *water-repellent preservative stain* or *exterior wood stain*. There are many types available, but all consist of resin solutions of low viscosity and low solids content: these solutions are readily absorbed into the surface layers of the timber. Their protective action is due in part to the effectiveness of water-repellent resins in preventing water ingress, and in part from the presence of finely dispersed pigments which protect against photochemical attack. The higher the concentration of pigments the greater the protection, but this is achieved at the expense of loss in transparency of the finish. Easy to apply and maintain these thin films, however, offer little resistance to the transmission of water vapour into and out of the timber. Compared with a paint or varnish the water-repellent finish will allow timber to wet up and dry out at a much faster rate, thereby eliminating problems of water accumulation which can occur behind impermeable paint systems; the presence of a preservative constituent reduces the possibility of fungal development during periods of high moisture uptake. The films do require, however, more frequent maintenance, but nevertheless have become well established for the treatment of cladding and hardwood joinery.

By far the most widely used finish, especially for external softwood joinery, is the traditional opaque alkyd gloss or flat paint system embracing appropriate undercoats and primers; a three- or four-coat system is usually recommended. Multiple coats of oil-based paint are effective barriers to the movement of liquid and vapour water; however, breaks in the continuity of the film after relatively short exposure constitute a ready means of entry of moisture after which the surrounding, intact film will act as a barrier to moisture escape, thereby increasing the likelihood of fungal attack. The effectiveness of the paint system is determined to a considerable extent by the quality of the primer. Quite frequently window and door joinery with only a priming coat is left exposed on building sites for long periods of time. Most primers are permeable to water, are low in elasticity and rapidly disintegrate owing to stresses set up in the wet timber; it is therefore essential that only a high quality of primer is used. Emulsion-based primer/undercoats applied in two consecutive coats are more flexible and potentially more durable than the traditional resin-based primers and undercoats.

A new range of exterior quality paints has been produced since the late 1980s; these are either solvent-borne or water-borne formulations. Some of the formulations have a higher level of moisture permeability than conventional paint systems and have been described as *microporous*. These are claimed to resist the passage of liquid water, but to allow the passage of water vapour, thereby allowing the timber to dry out; however, there appears to be no conclusive proof for such claims.

The solvent-borne exterior paints come in many forms, for example, as a three-layer system based on flexible alkyd resins which produce a gloss finish, or a one-can system which is applied in two coats and which produces a low-sheen finish.

The water borne exterior paints are based on acrylic or alkyd–acrylic emulsions applied in either two- or three-coat systems. The water-borne system has a higher level of permeability compared with the solvent-borne system. Even more important is the high level of film extensibility of the water-borne system which is retained in ageing (Miller and Boxall, 1994), and which contributes to its better performance on site than the solvent-borne exterior paints.

Test work has indicated that the pretreatment of surfaces to be coated with a water-repellent preservative solution has a most beneficial effect in extending the life of the complete system, first by increasing the stability of the wood surface, thereby reducing the stresses set up on exposure, and second by increasing adhesion between the timber surface and the coating. This concept of an integrated system of protection employing preservation and coating, though new for timber, has long been established in certain other materials; thus, it is common practice prior to the coating of metal to degrease the surface to improve adhesion.

One specialised group of finishes for timber and timber products is that of the flame-retardant coatings. These coatings, designed only to reduce the spread of flame, must be applied fairly thickly and must neither be damaged in subsequent installation and usage of the material, nor their effect negated by the application of unsuitable coverings. Nearly all the flame retardants on the UK market intumesce on heating and the resulting foam forms a protective layer of resistant char.

References

Standards and specifications

BS 1282 (1975) *Guide to the choice, use, and application of wood preservatives*, BSI, London.

BS 5707 *Solutions of wood preservatives in organic solvents* Part 1 (1979) – *Specification for solutions for general purpose applications, including timber that is to be painted.* Part 2 (1986) *Solutions of wood preservatives in organic solvents – Specification for pentachlorophenol wood preservative solution for use on timber that is not required to be painted.* BSI, London.

EN 300 (1997) *OSB – Definition and specification.*

EN 312–1 (1997) *Particleboards – Specifications – Part 1: General requirements for all board types.*

EN 312–2 (1997) *Particleboards – Specifications – Part 2: Requirements for general purpose boards for use in dry conditions.*

EN 312–3 (1997) *Particleboards – Specifications – Part 3: Requirements for boards for interior fitments for use in dry conditions.*

EN 312–4 (1997) *Particleboards – Specifications – Part 4: Requirements for load bearing boards for use in dry conditions.*

EN 312–5 (1997) *Particleboards – Specifications – Part 5: Requirements for load bearing boards for use in humid conditions.*

EN 312–6 (1997) *Particleboards – Specifications – Part 6: Requirements for heavy duty load bearing boards for use in dry conditions.*

EN 312–7 (1997) *Particleboards – Specifications – Part 7: Requirements for heavy duty load bearing boards for use in humid conditions.*
EN 314–1 (1993) *Plywood – Bonding quality – Part 1: Test methods.*
EN 314–2 (1993) *Plywood – Bonding quality – Part 2: Requirements.*
EN 335–1 (1992) *Hazard classes of wood and wood-based products against biological attack – Part 1: Classification of hazard classes.*
EN 350–2 (1994) *Durability of wood and wood-based products – Natural durability of solid wood – Part 2: Guide to natural durability and treatability of selected wood species of importance in Europe.*
EN 622–1 (1997) *Fibreboards – Specifications – Part 1: General requirements.*
EN 622–2 (1997) *Fibreboards – Specifications – Part 2: Requirements for hardboards.*
EN 622–3 (1997) *Fibreboards – Specifications – Part 3: Requirements for medium boards.*
EN 622–4 (1997) *Fibreboards – Specifications – Part 4: Requirements for softboards.*
EN 622–5 (1997) *Fibreboards – Specifications – Part 5: Requirements for dry process boards.*
EN 634–1 (1995) *Cement bonded particleboards – Specifications – Part 1: General requirements.*
EN 634–2 (1997) *Cement bonded particleboards – Specifications – Part 2: Requirements for OPC-bonded particleboards for use in dry, humid, and exterior conditions.*
EN 636–1 (1997) *Plywood – Specifications – Part 1: Requirements for plywood for use in dry conditions.*
EN 636–2 (1997) *Plywood – Specifications – Part 2: Requirements for plywood for use in humid conditions.*
EN 636–3 (1997) *Plywood – Specifications – Part 3: Requirements for plywood for use in exterior conditions.*
EN 789 (1996) *Timber structures – Test methods – Determination of mechanical properties of wood-based panels.*
EN 1058 (1996) *Wood-based panels – Determination of characteristic values of mechanical properties and density.*
ENV 1156 (1999) *Wood-based panels – Determination of load and creep factors.*
ENV 1995–1–1 (1994) Eurocode 5 *Design of timber structures – Part 1–1: General rules and rules for buildings.*
pr EN 12369 (to be published 2000/2001) *Wood-based panels – Characteristic values for use in structural design.*

Literature

Chou, C.K., Chandler, J.A. and Preston, R.D. (1973) Microdistribution of metal elements in wood impregnated with a copper-chrome-arsenic preservative as determined by analytical electron microscopy. *Wood Sci. Technol.*, **7**, 151–160.
Hill, C.A.S. and Jones, D. (1996) The dimensional stabilisation of Corsican pine sapwood by reaction with carboxylic acid anhydrides, *Holzforschung*, **50** (5), 457–462.
Hon, D.N.S. (Ed.) (1996) *Chemical modification of lignocellulosic materials*, Marcel Dekker, New York.
Miller, E.R. and Boxall, J. (1994) *Water-borne coatings for exterior wood*, Building Research Establishment Information Paper IP 4/94.

Morén, R.E. (1964) Some practical applications of polyethylene glycol for the stabilisation and preservation of wood, paper presented to the British Wood Preserving Association annual convention.

Rowell, R.M. (1984) Chemical modification of wood. *For. Prod. Abstracts*, **6**, 75–78.

Turner, P. and Murphy, R.J. (1998). Treatments of timber products with gaseous borate esters. Part 2: Process improvement. *Wood Sci. Technol.*, **32**, 25–31.

Further reading

Chapters 1–3 Wood structure and its variability; wood chemistry; density; appearance

Archer, R.R. (1986) *Growth Stresses and Strains in Trees*, Springer-Verlag, Berlin.

Desch, H.E. and Dinwoodie, J.M. (1996) *Timber – Structure, Properties, Conversion and Use*, 7th edn, Macmillan, Basingstoke, England.

Harris, J.M. (1989) *Spiral Grain and Wave Phenomenon in Wood Formation*, Springer-Verlag, Berlin.

Hillis, W.E. (1987) *Heartwood and Tree Exudates*, Springer-Verlag, Berlin.

Jane, F.W. (1970) *The Structure of Wood*, 2nd edn, Adam & Charles Black, London.

Preston, R.D. (1974) *The Physical Biology of Plant Cell Walls*, Chapman and Hall, London.

Wise, L.E. and Jahn, E.C. (1952) *Wood Chemistry*, 2nd edn, Rheinhold, New York.

Chapter 4 Moisture; shrinkage; movement; sorption; thermal expansion

Hoffmeyer, P. (Ed.) (1997) *International Conference on Wood-water Relations*. Proceedings of conference held in Copenhagen in 1997, published by the management committee of EC COST Action E8.

Kollmann, F.F.P. and Côté, W.A. (1968) *Principles of Wood Science and Technology, I, Solid wood*, Springer-Verlag, Berlin.

Siau, J.F. (1984) *Transport Processes in Wood*, Springer-Verlag, Berlin.

Skaar, C. (1988) *Wood-water Relations*, Springer-Verlag, Berlin.

Chapter 5 Flow in wood – fluid, thermal and electrical; permeability; diffusion

Hoffmeyer, P. (Ed.) (1997) *International Conference on Wood-water Relations*. Proceedings of conference held in Copenhagen in 1997, published by the management committee of EC COST Action E8.

Leyton, L. (1975) *Fluid Behaviour in Biological Systems*, Clarendon, Oxford.

Siau, J.F. (1984) *Transport Processes in Wood*, Springer-Verlag, Berlin.

Skaar, C. (1988) *Wood-water Relations*, Springer-Verlag, Berlin.

Torgovnikov, G.I. (1993) *Dialectric Properties of Wood and Wood-based Materials*, Springer-Verlag, Berlin.

Chapter 6 Deformation (mechanical); elasticity; viscoelasticity; creep; mechano-sorptive behaviour

Aicher, S. (Ed.) (1996) *International Conference on Wood Mechanics*. Proceedings of conference held in Stuttgart, May 1996, and published by the management committee of EC COST Action E8.

Bodig, J. and Jayne, B.A. (1982) *Mechanics of Wood and Wood Composites*. Van Nostrand Reinhold, New York.

Bonfield, P.W., Dinwoodie, J.M. and Mundy, J.S. (Eds) (1995) *Workshop on Mechanical Properties of Panel Products*. Proceedings of workshop held at Watford, 1995, and published by the management committee of EC COST Action 508.

Hearmon, R.F.S. (1961) *An Introduction to Applied Anisotropic Elasticity*, Oxford University Press.

Morlier, P. (Ed.) (1994) *Creep in Timber Structures*, Rilem report 8, E & FN Spon, London.

Chapter 7 Strength; duration-of-load; toughness; fracture mechanics; failure; fracture morphology

Aicher, S. (Ed.) (1996) *International Conference on Wood Mechanics*. Proceedings of conference held in Stuttgart May 1996, and published by the management committee of EC COST Action E8.

Barrett, J.D. and Foschi, R.O. (Eds) (1979) *Proceedings of the first international conference on wood fracture*, Banff, Alberta, 1978, Forintek Canada Corp.

Barrett, J.D. Foschi, R.O., Vokey, H.P. and Varoglu, E. (Eds) (1986) *Proc. International Workshop on Duration of Load in Lumber and Wood Products*, held at Richmond, BC, Canada, 1985, special publication SP-27, Forintek Canada Corp.

Bodig, J. and Jayne, B.A. (1982) *Mechanics of Wood and Wood Composites*. Van Nostrand Reinhold, New York.

Bonfield, P.W., Dinwoodie, J.M. and Mundy, J.S. (Eds) (1995) *Workshop on Mechanical Properties of Panel Products*. Proceedings of workshop held at Watford, 1995, and published by the management committee of EC COST Action 508.

Gordon, J.E. (1976) *The New Science of Strong Materials*, 2nd edn, Penguin.

Gowda, S.S. (Ed.) (1994) *Workshop on Service Life Assessment of Wooden Structures with Special Emphasis on the Effect of Load Duration in Various Environments*. Proceedings of workshop held at Espoo, Finland, May 1994, and published by the management committee of EC COST Action 508.

Morlier, P., Valentin, G. and Seoane, I. (Eds) (1992) *Workshop on Fracture Mechanics in Wooden Materials*. Proceedings of workshop held at Bordeaux, France, April 1992, and published by the management committee of EC COST Action 508.

Chapter 8 Durability; degradation

Bravery, A.F., Berry, R.W., Carey, J.K. and Cooper, D.E. (1987) *Recognising Wood Rot and Insect Damage in Buildings*, Building Research Establishment Report, BRE, Watford, England.

Chapter 9 Processing

Bonfield, P.W., Dinwoodie, J.M. and Mundy, J.S. (Eds) (1995) *Workshop on Mechanical Properties of Panel Products*. Proceedings of workshop held at Watford, 1995, and published by the management committee of EC COST Action 508.

Kollmann, F.F.P., Kuenzi, E.W. and Stamm, A.J. (1975) *Principles of Wood Science and Technology. Part II: Wood-based materials*, Springer-Verlag, Berlin.

Pizzi, A. (1983) *Wood Adhesives; Chemistry and Technology*, Marcel Dekker, New York.

Index